The Science of Soccer

Barbara Guth Worlds of Wonder

Science Series for Young Readers

Advisory Editors: David Holtby and Karen Taschek

*Also available in the University of New Mexico Press
Barbara Guth Worlds of Wonder Science Series for Young Readers*

THE SCIENCE OF SOCCER

A BOUNCING BALL AND A BANANA KICK

JOHN TAYLOR

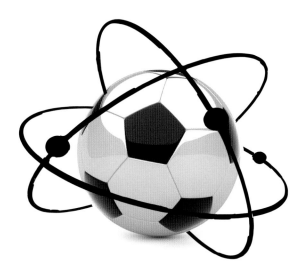

University of New Mexico Press | Albuquerque

19 18 17 16 15 14   1 2 3 4 5 6

LIBRARY OF CONGRESS CATALOGING-IN-PUBLICATION DATA

Taylor, John, 1947– author.
The science of soccer : a bouncing ball and a banana kick / John Taylor.
pages cm — (Barbara Guth Worlds of Wonder Science Series for Young Readers)
Includes bibliographical references and index.
ISBN 978-0-8263-5464-8 (cloth : alk. paper) — ISBN 978-0-8263-5465-5 (electronic)
1. Physics. 2. Soccer—Mathematics. I. Title.
QC26.T39 2014
796.33401'53—dc23
2013039145

Designed by Lila Sanchez
Set in Warnock Pro 11.5/18

Contents

Acknowledgments

Several individuals have encouraged and assisted in the production of this book. First, I would like to thank Dr. David Holtby from the University of New Mexico Press who invited me to start the project and encouraged me as I went along. After David's retirement, the project was carried forward by John Byram, whose enthusiasm and encouragement I much appreciate. The initial technical review by Dr. Steven Schafer of Sandia National Laboratories and the subsequent assessments by Dr. Clifton Murray of the University of New Mexico and Dr. Jay Williams of Virginia Tech were invaluable. Reviews by Eric Brown of Los Lunas Middle School and his students, Maryann Hospelhorn of Valencia Middle School and her students, and Peter Sawyer of The Thacher School helped me target the material for the intended audience. Faculty members Cathy Bailey, Kari Daniels, Klaus Weber, Michael VeSeart, Ivy Graham-Dewers, Chris Hannaford, and Beth Scanlon of the Bosque School and Bosque School students Elijah Martinez, Adrian Aleixandre, Ellie Fuerst, Jaclyn Pecille, and the Bosque School soccer teams actively participated in reviewing material, validating experiments, and staging and taking photos. A special debt of gratitude goes to Cathy Bailey, who spearheaded the effort at the Bosque School. Without her patience and enthusiasm for the project, it would not have been successfully completed. Finally, I gratefully acknowledge the love and support of my wife, Lynn, who unfailingly supported me and put up with mutterings about obscure physics terms for the duration of the project.

CHAPTER 1

The Beautiful Game

The score is tied with the championship on the line as a striker sprints toward the goal. As a pass comes in from the left wing, the striker is challenged in front of the penalty area by the last defender. The whistle blows and the referee indicates a direct free kick just outside the top of the penalty area. There are only moments left in this hard-fought game, and the referee will probably not add much time. The striker takes the ball from the referee and carefully places it just behind the arc at the top of the penalty area about 25 yards from the goal. The game and the season may ride on this kick. The keeper screams at the defenders to form a wall covering the far post. The keeper crouches forward just inside the near post as the referee signals for the kick to be taken.

The striker stands about 10 feet from the ball, looking through the wall toward the goal. As he accelerates toward the ball, he plants his right foot facing the keeper. His left foot swings in a giant arc, striking the ball slightly off center. The ball flies off at about 60 miles per hour on a trajectory just to the left of the wall. The keeper starts to relax—surely the ball will fade wide to his right and sail harmlessly across the end line. Suddenly, to his horror, the ball seems to snap back toward the goal. As he dives in desperation across the mouth of the goal, the ball curves into the net, just beyond his outstretched hands.

The whistle blows, the game is over, and the adulation of half of the crowd rains down on the field. Lost in the excitement is the subtle interplay of physics, biomechanics, kinesiology, and biochemistry that enabled the player to score this beautiful goal.

Soccer is a game of tactics and strategy with a rich overlay of history and tradition. It is, without question, the most popular sport in the world. In fact, the quadrennial World Cup tournament is the world's most-watched event, surpassing the Super Bowl, the World Series, and even the Olympic Games. But soccer is more than just a fascinating mobile chess match; it is also an endless scientific panorama. Each movement by every player and each interaction with the ball involves physics, fluid mechanics, biology, and physiology, to name just a few of the scientific disciplines.

This book explores the scientific aspects of the world's most popular sport. In general, we avoid the complex mathematics that generally accompany the physics, but enough mathematical background is provided to support the conclusions. For the curious, references are provided that include more detailed information.

The book is divided into six chapters, two appendices, and a glossary. The remainder of this chapter talks about the history of soccer from its earliest roots to the modern sport. It also takes a brief look at the history of soccer science. In chapter 2 we discuss some of the key physical and mathematical aspects of the game.

After establishing the historical and mathematical background, we examine the two key elements of the game—the ball and the players. Chapter 3 discusses how and why a ball bounces, and chapter 4 addresses how the ball spins and what that means for the game. In chapter 5 we look at the players kicking, heading, and trapping from a scientific point of view. We'll also look at the sources of the energy required to run, jump, and kick for an entire game.

Chapter 6 puts it all together by following a sequence of plays, not in the usual tactical way, but by describing the science behind tactical maneuvers. Sidebars and appendices allow those with a more mathematical bent to follow the physics and perform some experiments to see the effects of phenomena like drag, bounce, and spin. In addition, key terminology is highlighted, explained in the text, and summarized in the glossary. By the end of the book, we hope that you will have a better appreciation for the science that lies behind every move you make on a soccer field.

A Brief History of the Game

From ancient times, people have kicked objects across fields. Some of these games involved kicking and throwing, others just kicking. Some, like the Berber game of *koura*, the Japanese *kemari*, and the Aztec *tlachtli*, had religious or fertility overtones (in fact, the losers in *tlachtli* may have been sacrificed to the gods!). Some, like the Roman *harpastum* or the Chinese kicking game of *tsu chu* (or *cuju)*, were used to train soldiers for combat.

The Florentine game of *calcio* and English Shrove Tuesday games

Children in 12th-century China playing *tsu chu*   Episkyros

were more recreational, although violent, sometimes to the point of serious injury or even death. In fact, one medieval writer complained,

> Concerning football, I protest unto you that it may rather be called a friendly kind of fighting than recreation—sometimes their necks be broken, sometimes their backs or legs, sometimes their noses gush out with blood, and sometimes their eyes start out.

The ancient roots of soccer can be traced to Babylonia, China, and Egypt as early as 2500 BCE. The Babylonians played a game that involved a ball stuffed with bamboo fiber, sand, or hair, and there are scenes on Egyptian tomb walls of athletes juggling balls with their feet. The Greeks played a game called *episkyros* as long ago as 800 BCE, but this appears to have mostly involved throwing and catching. Another Greek game, *harpastron*, may have been more akin to modern rugby than to soccer, but it led to the Roman game of *harpastum*, a game that Julius Caesar and his legions brought to the various territories that they conquered across Europe. Although we know very little about the rules of *harpastum*, it probably included throwing and kicking a leather ball about eight inches in diameter and stuffed with chopped sponges or animal fur.

Football games pervaded the New World as well. Even before the European explorers first landed in North and South America, the Aztecs played *tlachtli*, the Inuits played *aqsaqtak*, and the Native Americans of eastern North America played *pusuckquakkohowog* (a word that means "they gather to play ball with the foot").

Julius Caesar's introduction of *harpastum* to the British Isles in 43 BCE would prove to be a critical point in the development of the game; if the roots of soccer's family tree spread into the ancient past, its trunk certainly grew strong and tall in England. The new game flourished in the British Isles, and there is a tradition that a group of Britons played a game at Derby on Shrove Tuesday (also known as Mardi Gras—the Tuesday before Ash Wednesday and the start of the Lenten season) in 217 CE to celebrate their military victory over a contingent of Romans.

Perhaps because they were played by soldiers, many of the ancient games came to be associated with commoners rather than the elite. This could also be the reason for their rather rough-and-tumble nature. Life in medieval Europe was downright painful much of the time, so any diversion was much appreciated. Soccer's medieval predecessors provided such a diversion, and the games became more and more popular. These early games had few if any rules—the guideline seems to have been "where the ball goes, the mob goes." *Soule* or *choule*, related to *harpastum*, sprung up in France. It was played using a leather ball stuffed with bran that was propelled with one's hands or feet or with sticks. The Florentine game of *calcio* had teams

Shrove Tuesday football

of 27 players, each attempting to get a ball through a narrow slit by kicking or throwing. Headbutting, punching, elbowing, and choking were permitted, but sucker-punching and kicks to the head were not allowed.

As the games grew in popularity among the commoners, they began to irk the monarchy. King Edward II banned the game in English cities in 1314, and King Richard reinforced the ban in 1389, complaining that playing football was interfering with archery practice. Under pressure, James IV reinstated the game in 1497, but Henry VIII banned it again in 1540. In 1597, Henry's daughter, Elizabeth I, levied a one-week jail sentence plus a church penance for anyone caught playing the game in London.

Royal attitudes changed as the game became more and more pervasive. By 1680, Charles II gave the game royal patronage, partially in reaction to the end of the Cromwellian revolution led by Puritans who frowned on games of any sort. As the British Navy spread the empire around the globe, soccer-like games went along—whenever a Royal Navy vessel docked in a new port, the sailors went ashore with their ball and the locals picked up the game.

The establishment of common rules for the game we now call soccer began in 1863 and had the effect of forcing those who wished to play games that involved handling or carrying the ball to set up their

British sailors playing soccer

Soccer's family tree

own organizations. In 1871, the Rugby Union was formed, named for Rugby, an English boys' school. In 1895, a splinter group formed the Rugby League, which played a game very similar to that of Rugby Union. In 1887, a game that combined elements of soccer and rugby that was popular in Ireland was formally included by the incorporation of the Gaelic Athletic Association (which also included an ancient Irish pastime called hurling, something of a cross between field hockey and lacrosse).

During the 1860s, a rugby-like game emerged in the United States under the influence of some Canadian rugby players. In the late 1860s and early 1870s, this game was being played at American colleges like Princeton, Harvard, and Rutgers and would eventually become American football. The game played at American colleges also "migrated" back to Canada and resulted in the game now known as Canadian football. Rugby also migrated to Australia where a game called Australian Rules football developed in the late 1870s.

Thus, by the late 19th century, the family tree of football had six main branches, four of which involved a non-spherical ball, and two of which retained the round ball. Only one of these games prohibited handling the ball, but this is the one that has grown to become the most popular sport in the world, formally known as Association Football, but known and loved by much of the world as soccer.

A Brief History of Soccer Science

What about the science of soccer? Soccer science is a relatively new discipline, but it has its roots in the studies of mechanics, gravity, and forces by early Greek thinkers like Aristotle, who first speculated about the physical world and how it worked. By laying out the principles of classical physics, people like Galileo Galilei and Sir Isaac Newton also contributed to soccer science.

Certainly, the players of the ancient ball games must have quickly figured out that the ball could be made to curve in flight, and they would have noticed that inflated balls bounced better than solid balls. However, it would take mathematical developments like calculus and engineering developments like fluid dynamics before the "science of soccer" could really emerge. In fact, although Sir Isaac Newton and others speculated on the spinning and curving behavior of tennis balls, the earliest serious studies of the behavior of objects as they flew through the air were by men like Heinrich Magnus and Daniel Bernoulli, who concerned

Aristotle

Daniel Bernoulli

John Strutt, Lord Rayleigh

Galen

themselves with the flight of cannonballs on the battlefield.

Toward the end of the 19th century, a Scottish mathematician named Peter Guthrie Tait performed a detailed analysis of the effect of the dimples on golf balls on their trajectory, and John Strutt, better known as Lord Rayleigh, followed in Newton's footsteps with a detailed analysis of the effect of spin on tennis balls.

As the popularity of soccer grew, the same principles that had been applied to tennis and golf were applied to "the beautiful game" in an effort to understand the physics of ball behavior.

The other aspect of "soccer science" has to do with the movement of players and how they interact with the ball. This field of study, known as **biomechanics** or **kinesiology**, traces its roots to Aristotle's "Parts of Animals," "Movement of Animals," and "Progression of

Animals," all written in the fourth century BCE. In fact, this Greek philosopher and scientist is sometimes referred to as the Father of Kinesiology.

The famed Roman physician Galen (131–201 CE) may have been the first "team physician" and is sometimes referred to as the Father of Sports Medicine by virtue of his work with gladiators. By studying their horrible wounds and nursing them back to fight another day, Galen came to understand the workings of muscle and bone better than anyone before him.

Leonardo Da Vinci was interested in the study of musculoskeletal movement and actions and described the mechanics of movement in the early 16th century. Major advances in the study of human anatomy and muscle anatomy plus the advent of motion pictures and stop-action photography has enabled modern-day kinesiologists to analyze movements like kicking, heading, and trapping in great detail.

CHAPTER 2

An Introduction to the Physics and Mathematics of Soccer

Although this book is not a math or physics textbook, there are a few basic concepts that we need to introduce in order to establish a common vocabulary and understanding. First, let's talk about vectors. A **vector** is a way of describing a quantity in terms of its size (called magnitude) and its direction. For example, a moving soccer ball is going in a specific direction at a specific speed. The speed and direction can be described in terms of a vector—the direction is simply the direction it is moving and the magnitude is how fast it is moving. Similarly, when you kick or head a ball, you exert a **force** on the ball that has a specific direction (the direction that your head or foot is moving relative to the ball) and a magnitude (the amount of force you are applying—i.e., how hard you are kicking or heading the ball).[*]

Velocity and acceleration are two important aspects of motion that also need to be defined. Both are vectors because they have magnitude and direction. **Velocity** is the amount of distance covered by a ball divided by the time required to cover the distance (like the speed of the ball), together with its direction of motion. Velocity is

---

[*] Throughout the book I will take some liberties with pure physics definitions. For example, I will treat mass and weight as equivalent.

measured in units like meters per second (m/s).* **Acceleration** is a measure of how fast the ball is either speeding up or slowing down, together with the direction it is going as it changes speed. In other words, acceleration is the change in velocity in a specific period of time. An example of acceleration is a ball that is kicked from a fixed position like a goal kick, a corner kick, or a free kick. In the short time that the foot is in contact with the ball (.005 to .01 seconds), it is accelerated from zero speed (when it is sitting on the field and not moving) to full speed when it leaves the foot. Similarly, as soon as the foot is no longer in contact with the ball, it begins to slow down because of friction with the surrounding air. Acceleration is a vector because it has a magnitude (how fast the speed of the ball is changing) and a direction (the direction of the foot as it accelerates the ball).

We can divide the vector that describes the initial velocity of the kicked ball into two pieces, or **vector components**, one that moves parallel to the surface of the field (the horizontal component) and one that moves perpendicular to the field (the vertical component). Therefore, we can diagram the velocity vector that represents the initial movement of the ball into a right triangle—the base of the triangle describes the movement of the ball parallel with the field, and the height of the triangle describes the ball's vertical movement. Separating the movement of the ball in this way makes it easier to explain why the ball follows a particular path.

Another basic concept is **energy**. While energy is a highly useful concept, it is so basic that it is not easy to define. However, there are formulas for the various types of energy, and we can also illustrate them by using examples. Three forms of particular interest to soccer players are **energy of motion** (kicking, heading, throwing), **energy of position** (height above the ground), and **metabolic energy** (energy obtained from the food we eat).

---

* This book will use metric units with their English unit equivalents in parentheses where appropriate. Note that if you don't care about direction, velocity and speed are equivalent.

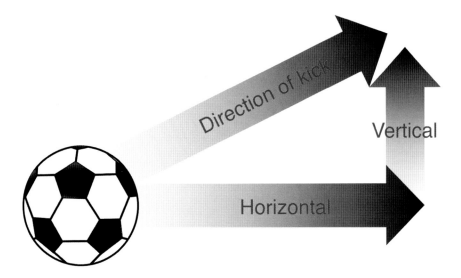

When someone or something causes an object to change speed or direction, we can say that a force has acted on the object. If this force acts over a distance, we define the product of the force and the distance as **work**. Mathematically, work and energy are equivalent. For our purposes, we can consider that energy either makes things change or has the potential to make things change. While work relates the force and the distance over which the force acts, **impulse** is the product of the force and the time during which the force is applied.

For example, if a ball is stationary and you run up and kick it, your leg and foot move rapidly toward the ball, eventually hitting it and causing it to move. This movement can be described in terms of the velocity of your foot (remember that velocity is a vector and includes both speed and direction) when it hits the ball and the velocity (speed and direction) of the ball afterward. The energy of this movement is called **kinetic energy**.* The force exerted by your foot while it is in contact with the ball acts over the distance during which there is contact between your foot and the ball (typically less than about 30 cm). The force of your foot against the ball times the distance

---

* Mathematically, kinetic energy is equal to half the mass of the moving object times its velocity squared ($KE = \frac{1}{2} mv^2$).

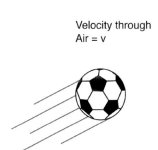

Velocity through
Air = v

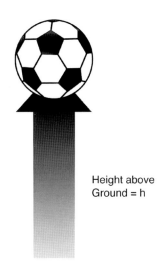

Height above
Ground = h

Kinetic energy = energy of motion = ½ × mass × (velocity)²
Potential energy = energy of vertical position = mass × acceleration of gravity × height

during which there is contact is the work done on the ball, and the force times the length of time your foot is in contact with the ball is the impulse applied to the ball. The movement of your foot when it hits the ball has caused a change in the movement of the ball—therefore, there has been a transfer of the property we call energy. The energy of the ball equals the work done on it by the foot. This transfer of energy is obvious—the ball speeds up and your foot slows down. Another important concept of motion, momentum, will be discussed shortly.

If you pick up a ball and drop it, it accelerates downward as it is pulled by the earth's gravity. When you lifted the ball, you moved its mass in an opposite direction to the force of gravity. Since you have moved the ball upward against a force (gravity), the work done on the ball is equal to its mass times the distance you lifted it times the force of gravity. That is why you feel the weight of the ball as you lift it. All the time you were holding and lifting the ball, it had the potential to accelerate downward. Therefore, the energy which is associated with height but not motion is called **potential energy.**[*]

---

[*] The mathematical expression for potential energy is the product of the mass of the object times the acceleration of gravity—9.8 m/s² (32.2 ft/s²) times the height to which the object is lifted. For our purposes, we will consider that weight and mass are the same.

Sir Isaac Newton

In order to understand why the soccer ball moves in the first place, we have to turn back the calendar to 1687, when Sir Isaac Newton first described the three laws of motion that govern the movement of all objects, from planets to soccer balls.

Newton's first law says that an object will continue doing what it is doing unless it is acted on by a force like gravity, pressure, or the force caused by striking a foot or head or goalpost. Said another way, objects tend to continue moving unless something acts to change their movement. If you roll a soccer ball along a very smooth surface, it will simply continue to roll until it hits something like a wall or your foot.

Newton's second law defines "force" in terms of measurable quantities. We know the mass of a soccer ball, and if we have the right tools, we can measure its movement after being kicked or headed. The product of mass and acceleration defines the force that causes the change in movement (F = ma). Recall that acceleration is a vector—that is, it has both magnitude and direction. Therefore, force is a vector—it acts in a specific direction with a specific magnitude.

Newton's third law says that if a force is exerted on a soccer ball by your head or your foot, the soccer ball exerts an equal and opposite force on your head or your foot. This is the same as saying that when you move your foot to kick a ball that is sitting on the ground, it is the same as if your foot was on the ground and was hit by a ball with the same force. Clearly, the mass of your foot and leg

Force of the foot on the ball is equal and opposite to the force of the ball on the foot.

are different from the mass of the ball. In addition, the acceleration of your leg and foot will be different from the acceleration of the ball. But the product of the mass of your foot and leg times the acceleration of your foot and leg is equal and opposite to the product of the mass of the ball and its acceleration.

A soccer game can be viewed as simply a sequence of applications of Newton's three laws. At the kickoff the ball is at rest in the center of the field. A striker kicks it, imparting a force which, by the second law, is equal to the mass of the ball times its acceleration. After the force has acted on the ball, it moves in accordance with the first law. However, just as soon as it starts to move, forces like aerodynamic drag, friction with the grass, and gravity begin to act on it, so its motion is altered again. And so on and so on throughout the game.

Similarly, when a ball that has been kicked into the air returns to the ground or to your head or foot, it is deformed as it strikes the field/head/foot and is subject to pressure forces inside the ball itself, causing it to bounce.

Forces can be divided into two categories—**contact forces** and **non-contact forces**. Gravity is an example of a non-contact force because the ball is separate from the earth, yet there is an attraction between the two that causes the ball to accelerate toward the earth.

Kicking, friction with the grass, and air resistance (also called aerodynamic drag) are all contact forces—they require that two objects (your foot and the ball, the ball and the grass, the ball and the molecules in the air) be in contact for the force to act.

You may have learned that only four types of forces have been discovered to date—the gravitational force, the electromagnetic force, and two forces that only exist in the nuclei of atoms and that need not concern us here. So, what is this about contact forces? If you have ever seen a "diagram" of an atom like the one on the right, you will note that it is mostly empty space with a cloud of electrons buzzing around the nucleus. This can be thought of as an arrangement like our solar system, which is also mostly empty space with the sun at the center and the planets rotating in orbits a long way from the sun.

So, if atoms and molecules consist mostly of empty space, how can we have a "contact" force? The answer is that there really never is "contact" between objects. If you have bumped your head on a goalpost or been kicked in the shin, you may say, "Oh yes there is!"

However, what happens when two objects come very close to each other is that the electric fields of the molecules begin to interact.

In general, it is the electric fields of the electrons (the small blue balls in orbits around the nucleus in this diagram of a lithium atom) that interact. Similar charges (like negative and negative or positive and positive) repel each other; and electrons are negatively charged. Therefore, when the electrons in the atoms and molecules of your shoe get closer and closer to the electrons in the atoms and molecules of the ball, their respective electrons repel each other. The closer they get, the stronger is the repulsion. It is this repulsive force by the electric fields in the trillions and trillions of electrons in your shoe and the ball that results in the ball flying away from your shoe (or your head or shin hurting!). So, although the electrons never really touch each other, the separation distances are so small that, for our purposes, we will call these contact forces.

Newton's third law describes equal and opposite reaction forces. This can also be referred to as symmetry of forces. Now, let's introduce another version of symmetry, called **analytical symmetry**. Consider a soccer ball flying through the air. The principle of analytical

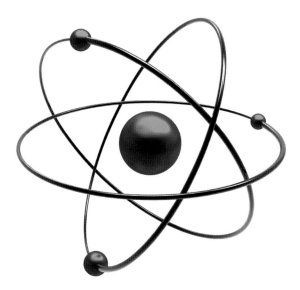

symmetry says that we will be able to understand the forces on the ball by either considering the ball to be moving through the air or by considering the ball to be stopped with the air moving past it.

It is also important to have a basic understanding of a principle known as **conservation of energy**. This principle says that energy can never be created or destroyed. Another way of saying this is that in any interaction, the total energy before and after the interaction is the same. While this may seem simple, it can get rather complicated. For example, a ball rolling across the field will eventually stop. A ball that is dropped on a field will bounce, but if it is left alone, eventually it will stop bouncing. Aren't these examples of energy being destroyed or somehow used up? Actually, they are not, but in order to explain this, we must talk about more than just the ball or the player; we must think of the entire set of balls, players, air, field, etc. that is involved.

For example, when a rolling ball slows down and stops, the energy that was involved in its rolling motion has simply been transferred to the grass (and a small amount to the surrounding air). If you could make extremely sensitive measurements, you would find that when the ball rolls over a blade of grass, the grass moves to a new position. Each blade of grass also rubs along the ball as it passes. Although the energy transferred from the ball to each blade of grass is small, there are a lot of blades of grass! Similarly, as the ball rolls along the field, the top part is moving through the air. Air molecules are pushed by the rolling ball and this pushing transfers energy from the rolling ball to the air molecules.

You can make a similar detailed evaluation of a bouncing ball, except in this case the energy losses include not only the deformation of the grass and movement of air molecules surrounding the ball, but also the movement of the air molecules inside the ball and the molecules of the ball itself. Each time the ball bounces, some energy is transferred to the air inside, making the molecules move faster.

With a sensitive instrument, you could actually measure this effect as heat in the air. A similar phenomenon occurs in the material in the ball itself, also resulting in heat. Thus, the ball's energy of motion has been converted to heat, both of the ball and of the air inside it. We will talk more about this in the next chapter.

So, we can say that energy is never created or destroyed, although it can change form.

Another important principle is **conservation of momentum**. **Momentum** describes the tendency of an object to continue moving at a constant speed in a constant direction unless it is acted on by a force, in accordance with Newton's first law.[*] Now consider that we have an interaction between two objects, say two balls on a pool table. If you hit one ball straight on with the cue ball, the cue ball will stop and the other ball will move with the same speed that the cue ball had when the two collided. We say that the momentum of the cue ball has been transferred to the other ball so that the total momentum of the two balls prior to the collision (one ball at rest with zero momentum and the cue ball moving along the table with some momentum) has been kept constant. In the jargon of physics, momentum has been conserved.

When we examine how forces act on an object, we also need to introduce a concept known as **center of mass**. When you kick a soccer ball, the force is applied to a small area on the surface of the ball. However, the entire ball, not just the small part of the ball that is in contact with the foot, will respond to the kick. By analyzing how the force at one point is distributed across the entire surface of the ball, it can be shown that the ball will behave as if an equivalent force had been applied at a single point on the ball. We call this point the center of mass. The center of mass of a soccer ball is almost always at the center of the ball.

----

[*] Mathematically, momentum is equal to the product of mass and velocity (M = mv).

If the vector that describes the movement of the foot prior to and during the kick passes directly from the point of impact on the surface of the ball through the center of the ball, the initial velocity vector of the ball after the kick will follow the velocity vector of the foot. However, if the vector that describes the movement of the foot does not pass directly through the center of the ball, we must divide the force vector of the foot into two components. Unlike our previous example in which one component was parallel to the field and one perpendicular, in this case one component passes through the center of the ball, but its perpendicular partner does not. This leads us into a discussion of torque.

You know that if you kick a ball at dead center, it does not spin, but if you kick it above or below or to the right or left of the center, it not only moves forward but also has some spin. This is the result of **torque**. If you kick the ball off center, you have applied a force away from the center of mass of the ball. This is the same as hitting a slice in tennis. The distance from the center of the ball to the line of direction of the kick is called the **moment arm** or torque arm—the longer the moment arm (for the same strength of kick), the more the ball will spin. We will spend a lot of time talking about spin and its consequences later on.

Every kick causes the ball to move through the air, even if the ball is just rolling along the ground. The study of the movement of objects through air is called **aerodynamics** (a subset of a larger field of study called **fluid mechanics**). Here we will use analytic symmetry to say that the way a ball moves through the air at a specific velocity is the same as the way air moves around a stationary ball at the same velocity.

Consider air blowing toward a stationary ball. This air contains molecules that can be described in terms of speed, direction, temperature, and pressure. All of these factors combine to give the air a

torque (τ) = Force × moment arm ($F = \tau d$)

characteristic energy. Since this energy can neither be created nor destroyed, the sum of the energies of the air and the ball before the air reaches the ball must be the same as the sum of the energies during the period of time that the air is flowing around the ball and must be the same after the air has passed the ball.

Now, think of this air being three meters (10 feet) in front of the ball and moving to a position three meters behind the ball, passing over the surface of the ball as it moves. The air above or below the ball can move the six meters in a straight line, but any air that is aimed at the ball must move around the ball. The air that moves around the ball must travel a little farther since it must curve around the ball rather than move in a straight line. Since all the air was together before it came to the ball, it must get back together after passing the ball. Therefore, as the air moves around the ball it accelerates, and when it returns to its previous position after passing the ball, it decelerates.

Because energy must be conserved, the increase in speed of the air nearest the surface of the ball (which increases the kinetic or

Experiment—*Conservation of energy in fluid flow*

Purpose This experiment demonstrates conservation of energy in a fluid like air.

Equipment required A large fan, a beach ball, a child's pinwheel and/or a handheld anemometer (an instrument that measures wind speed)

Experimental procedure Place the fan horizontally across two chairs or two sawhorses so that the airflow is vertical rather than horizontal. Turn the fan on and place the inflated beach ball in the airstream. Try to move the ball out of the airstream by gently tapping it. If the fan is equipped with a variable speed control, vary the speed and graph the height of the ball as a function of the speed of the airflow.

Questions to be answered in this experiment
1. Why does the beach ball remain in the airstream?
2. What forces are acting on the beach ball?
3. What is the effect of varying the speed of the airflow?

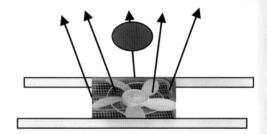

(See Appendix A, Experiment #3 on page 73 for further discussion.)

movement energy of those air molecules) must be accompanied by a decrease in some other form of energy, in this case the pressure of the air molecules. This trade-off between the speed of the air and its pressure is called the **Bernoulli effect** and accounts for aerodynamic effects like the lift that enables airplanes to fly.

In our case, however, the decrease in the pressure force will be uniform over the entire ball because the air is moving both above and below it. If this were all that there was to the aerodynamics of a soccer ball, our "beautiful game" would be a rather boring one. However, there is an additional factor that we must consider that combines the flow of the air over the ball with the effects of friction between the air molecules and the surface of the ball.

Recall our discussion of the electrical interactions between the electrons in atoms and molecules that come very close together. This also happens when a moving ball passes through the air. As the ball moves through the air, molecules of air near the surface rub against the ball, transferring energy from the ball to the air, causing the air to speed up and the ball to slow down. This phenomenon is called **aerodynamic drag**. The air molecules also rub against each other. This rubbing action slows the molecules down as their kinetic energy is converted to heat.

Right at the surface of the ball the air is moving rather slowly because some of the air molecules actually "stick" to the surface of the ball due to interactions between the electrons on the surface of the ball and electrons in the air molecules. In this case, the actions are attractive rather than repulsive because the materials tend to try to "share" electrons.

This "electron sharing" creates a layer of air around the ball that is called the **boundary layer**. Because this layer is very thin (a millimeter or so on a fast-moving soccer ball), small differences in the roughness of the surface of the ball caused by stitching, joints, or scratches can interact with the air, causing the ball to speed up or slow down even more.

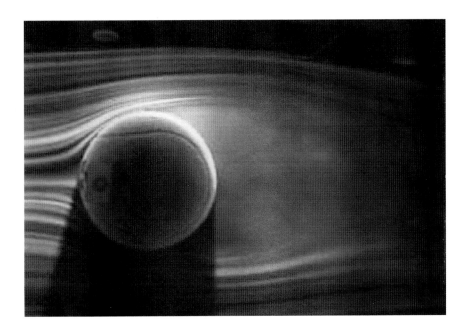

As the ball moves faster and faster, the boundary layer gets thinner and thinner. If you increase the speed even further, the flow actually separates from the surface.* With no spin this separation occurs near the back of the ball, leaving very disturbed air moving in all directions behind the ball as it travels through the air. This disturbance of the air is called **turbulence**. In general, areas of high turbulence have a lower pressure than areas of lower turbulence. When the surface of the ball is very rough (for example, the furry surface of a tennis ball or the dimpled surface of a golf ball), the separation occurs at lower speeds. For smoother balls, like soccer balls, this separation occurs at higher speeds. As we delve more deeply into plays like the inswinging kick, we will see why this is important.

---

* This does not mean that there is no air near the ball, only that the smoothly flowing airstream and the boundary layer around the ball have separated from the surface.

CHAPTER 3

Follow the Bouncing Ball

Some ancient ball games were played with balls that were animal skins stuffed with hay or sand. There are also stories of games played with the heads of defeated enemies! These games must have been a bit slow-paced, since some of the most intriguing aspects of the game occur because the ball bounces. After all, heading or kicking a ball filled with sand or straw (or a recently decapitated head!) would be both painful and frustrating. Perhaps this is why many of the earliest ball games involved juggling or catching the ball and throwing it.

Although there is some evidence that versions of the Chinese game of *tsu chu* may have used inflated animal bladders, we know for sure that inflated balls, perhaps animal bladders encased in leather, were used in 16th-century Italy. Some of the Central American games may have used solid rubber balls. Both of these cases provided a ball that behaved "elastically" as described below. This changed the game forever, providing for higher speeds as well as bounce and spin that were simply not possible with the heavy, solid balls used in some of the ancient games. So why does a ball bounce? Well, there are basically two phenomena at work—**elasticity** and **compression**.

Let's first consider elasticity. The materials in which we are interested are composed of molecules that, in turn, are composed of atoms.

Inflating a ball in 16th-century Italy

The molecules are held together by electric forces in which electrons are shared between atoms. Most of these molecules have a preferred shape that minimizes the amount of energy required for them to stay in place. Shown on page 29 is a molecule of adenosine triphosphate, one of the chemicals that is critical for energy production in the body.

When an external force acts to deform these materials, it moves the molecules away from their preferred position and their preferred shape. The mechanism and degree to which the material returns to its original shape and position can be described as elastic, plastic, or brittle behavior. **Brittle materials** do not deform easily, and when they do, they usually break (think about glass or bone). **Plastic materials** may deform rather easily, but do not return to their original shape (think about a dent in a car or a footprint on the field). **Elastic materials** return to their original shape after deformation (think about a soccer ball).

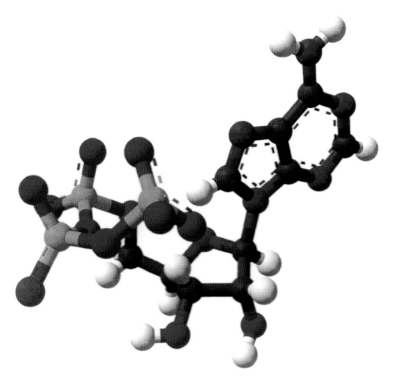

Adenosine triphospate (*Black*: Carbon, *White*: Hydrogen, *Red*: Oxygen, *Blue*: Nitrogen, *Orange*: Phosphorus)

A soccer ball generally consists of a rubber bladder surrounded by a protective case made of leather or some other material. The air inside the bladder is composed of molecules of various gases (nitrogen, oxygen, carbon dioxide, argon) moving around inside the fixed volume. The more molecules there are inside the volume, the more they bump into each other and the more they bump into the inside of the bladder. The sum of all of these collisions generates what we call **pressure** (the force on the inside of the ball divided by the area of the inside of the ball). The pressure inside the ball can be changed in a number of ways: the ball can burst, releasing the molecules; the coach can inflate the ball, adding more molecules to the already crowded volume; or the ball can strike a firm surface like the ground, the goalpost, or your foot. In this last case, the outside of the ball (both the casing and the rubber bladder) is pushed in on one side by the ground, the post, or the foot. This effectively reduces the volume

Soccer ball deforming against a solid object

of the ball. Since the number of molecules has not changed, the reduction in volume means that there will be more collisions of gas molecules with the inside of the ball, increasing the pressure. We can show that the product of pressure and volume is constant for a given number of molecules at a given temperature. This relationship is known as **Boyle's law**.

Now, let's put elasticity and compression together. Because of the choice of materials, rubber and leather, the casing and bladder of the ball deform elastically. As they deform, the volume inside the ball is decreased, resulting in a corresponding increase in pressure.

The increased pressure presses against the inside surface. Also, because the casing and bladder are elastic materials, the molecules in the leather case and rubber bladder are attempting to get back to their original shape. The net result is that the ball is pushed off the solid surface. Of course, if the surface is your head or foot or the goalkeeper's fist, even more energy may be imparted to the ball (the ball simply sees this as more deformation and increased internal pressure), and it may leave with an even greater velocity.

We lump the elastic deformation and forces that arise from

Experiment—*Follow the bouncing ball*

Purpose This experiment illustrates the behavior of a bouncing ball.

Equipment required Soccer ball, baseball, pool or croquet ball, inflation pump and pressure gauge, various surfaces (concrete, grass field, etc.), yardstick

Experimental procedure Drop each of the balls from a fixed height onto various surfaces and measure the height of the rebound. Vary the height from which the ball is dropped, the pressure in the ball, and the surfaces on which the ball is dropped. Make several measurements at each drop height, pressure, and surface. Graph the rebound height (*y* axis) as a function of the drop height (*x* axis).

Questions to be answered in this experiment
1. Which ball bounces highest? Why?
2. How does the height of bounce depend on the surface?
3. How does the height of bounce depend on the pressure of the ball?
4. Can you derive an expression for the coefficient of restitution that relates the ratio of the drop and rebound heights to the drop and rebound velocities?

(See Appendix A, Experiment #1 on page 65 for further discussion.)

increased internal pressure into a relationship that is called the **coefficient of restitution**. This relationship is defined as the velocity of the ball after striking a surface divided by the velocity of the ball before it strikes the surface. A typical coefficient of restitution of a properly inflated soccer ball bouncing off a concrete surface is about 0.8. This means that the velocity of the ball after striking the surface is 80 percent of the velocity before striking the surface. Because a grass field deforms plastically when struck by a ball, the coefficient of restitution of the ball on a field is less than that on a concrete surface. Since the coefficient of restitution for a soccer ball bouncing on a grass field is about 0.5, the height of the bounce will decrease with each successive bounce until the ball eventually comes to rest.

The relationship between the height of bounce for successive bounces is given by $h_2 = e^2 h_1$, where h_2 is the height after the bounce, e is the coefficient of restitution, and h_1 is the height before the first bounce. This expression is derived by defining the coefficient of restitution as the ratio of velocity after bounce to the velocity before bounce. Thus, with a coefficient of restitution of 0.5, each successive bounce is only 25 percent as high as the preceding bounce.

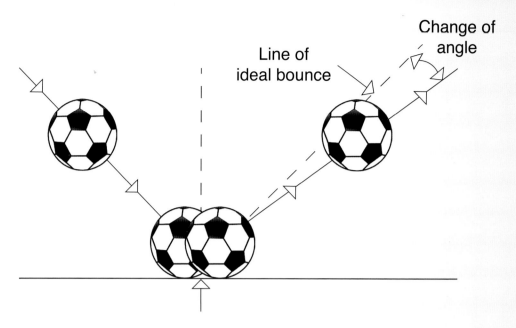

Line of
ideal bounce

Change of
angle

Ball sliding or skidding through bounce

Two other interactions of the ball with a surface like the field are rolling and skidding. As a ball moves along a surface, electrical forces are acting at a microscopic level, resulting in what we refer to as **friction**. In effect, whenever the ball rolls across a blade of grass, the grass tries to pull the ball backward because the electrons in the grass are being "shared" by the electrons in the surface of the ball. A ball rolling across a field will always slow down and will eventually stop unless it is kicked by a player.

Skidding can be viewed as a special case of rolling or bouncing. Each of us has had the experience of playing on a field after (or even during!) a rainstorm. Under these conditions, the goalkeeper must be particularly vigilant for balls that are bouncing right toward him because the ball may appear to jump forward off the wet grass. Typically, players miss the first few kicks or traps of the morning because the ball just doesn't go where it is supposed to go. This is the phenomenon of skidding. What has happened is that the field has been lubricated by the moisture. Materials that lubricate (think of the oil in a car engine) tend to have electrons that are not shared to the degree that those in the grass are, so water molecules come between the blades of grass and the casing of the ball and the grass is less able to pull on the ball and slow it down.

Thus, when the ball strikes a wet field, it loses less of its energy to friction with the grass. If it is rolling, it will roll farther and faster (assuming, of course, that the water is not actually puddling on the field!). These factors combine to make the ball slide as well as bounce or roll when it hits the field at an angle. This means that the rebound that would be expected on a dry field will be at a different speed and angle on a wet field.

So, we now have a basic understanding of how a ball behaves as it is bouncing, being kicked or headed, and rolling or skidding along the field. Next we will examine what happens as the ball moves through the air.

CHAPTER 4

How about a Banana Split?

One of the most fascinating aspects of ball games, from baseball to golf to tennis, is that players can make the ball change direction in flight. Golfers either cultivate or dread their slice; tennis players hit twisting serves that literally jump away from the receiving player; and baseball pitchers rely on curve balls, sliders, and knuckle balls.

Soccer, too, has its "curve balls," referred to as inswingers, out-swingers, or banana kicks. To understand why a soccer ball curves, we will need to answer several questions—what makes a ball spin? What are the effects of a spinning ball flying through the air? What happens if a ball flies through the air without spinning? As we address these phenomena, we will use the concepts of energy, momentum, mass, center of mass, friction, torque, and fluid flow.

First, let's talk about how a ball moves through the air. Because air is all around us, we don't think much about it until we see the strength of a hurricane, try to ride a bike into the wind, or try to kick a goal kick with the wind in our face. A ball that is kicked, headed, or thrown must push air molecules out of the way as it flies through the air. As the ball pushes on the molecules, some of its kinetic energy (energy of motion) is transferred to the air molecules. Since the kinetic energy of the ball is now lower, and since kinetic energy is equal to half the mass times the velocity squared (see chapter 2), the ball must slow down. However, since energy must be conserved, and since the kinetic energy has been transferred to the molecules, the local pressure near the ball increases. Molecules are pushed around the ball as it flies through the air. The air molecules speed up and the ball slows down.

The phenomenon of a ball slowing down as it moves through the air is called **aerodynamic drag**. The amount of drag is directly affected by the number of molecules that must be pushed aside. This number of molecules depends on the density of the air, the temperature, the humidity (the number of water molecules that are also in the air), and the altitude. For example, a ball kicked on a hot, dry summer day in Albuquerque or Denver will travel farther than one kicked with equal force on a cold, humid day in New Orleans or Miami because the air that must be pushed aside is less dense when it is hotter, drier, and higher in the atmosphere.

Because drag acts as a force slowing down the ball, it can be treated as a vector. Since it is proportional to the square of the velocity (Drag $\sim v^2$) and since horizontal velocity is usually much greater than the vertical velocity, the horizontal component of velocity is usually much more affected by drag than the vertical component.

Experiment—*Drag*

Purpose This experiment illustrates the drag felt by the ball when it passes through a medium like the atmosphere. We will use much denser materials like oil, water, and shampoo to show the effects of drag more dramatically.

Equipment required Glass bottle or test tube, marble, stopwatch, various liquids (like water, motor oil, shampoo, or syrup)

Experimental procedure Fill the glass container with one of the fluids. Release the marble into the container and measure the time it takes to fall to the bottom. One of these experiments should be done with an empty container (that is, a container filled only with air).

Questions to be answered in this experiment
1. What is the effect of drag on the movement of an object?
2. Does the effect of drag change with the material that the ball is passing through?

(See Appendix A, Experiment #2A on page 67 for further discussion.)

Drag is incorporated into mathematical expressions using a **drag coefficient**. For the ball speeds that we will encounter in a soccer game, this coefficient can be considered to be constant, so we can say that drag is directly proportional to the density of the air and the square of the velocity of the ball ($D \sim \rho v^2$).

Now that we have introduced drag, let's talk about how we make a ball spin. When a player kicks a ball, his or her foot follows a path to the ball and continues for the few milliseconds that the foot is in contact with the ball.

The foot's path also continues after the ball is no longer in contact (the so-called follow-through[*]). If the foot's projected path passes through the center of the ball (recall that the center of a uniform spherical object like a ball is also the center of mass of the object), the ball will fly straight off along an initial path defined by the point at which the foot strikes the ball (called the point of impact) and the

---

[*] The follow-through is important because it allows the leg and foot to come to a controlled stop after the energetic movement involved in the kick itself. In addition, kicking technique that includes a good follow-through can lengthen the period of contact with the ball, increasing the impulse and, therefore, the transfer of energy, and helping to direct the path of the ball with greater accuracy.

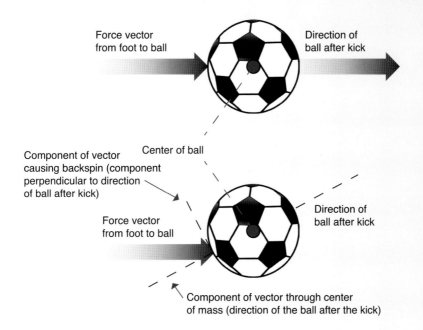

Force vector
from foot to ball

Direction of
ball after kick

Center of ball

Component of vector
causing backspin (component
perpendicular to direction
of ball after kick)

Force vector
from foot to ball

Direction of
ball after kick

Component of vector through center
of mass (direction of the ball after the kick)

center of the ball. The velocity vector of the ball (both speed and direction) will be determined by the force vector exerted by the foot on the ball (both magnitude and direction).

If the vector that defines the path of the foot does not pass through the center of the ball, we need to break the vector into two pieces—one that goes from the point of impact through the center of mass and one that is perpendicular to the original vector (or tangent to the surface of the ball). As an example of this sort of kick, let's consider a goal kick that is intended to go to strikers waiting near midfield. The ball must be lifted into the air, so the keeper strikes the ball below its center. The component of the force vector of the kick that passes through the center of mass of the ball determines the initial path. The perpendicular component, acting at the point of impact, exerts a torque on the ball, causing it to rotate backward (see figure on page 23).

Spin is also induced by the friction between the instep of the shoe (which usually has a high degree of roughness because of the laces) rubbing across the surface of the ball. Because of the electrostatic forces that exist on the molecular level, the surface of the ball is "pulled" in the direction of motion of the shoe as it passes across the surface, inducing spin. If the kicker maintains the same force for his or her kick but moves the point of impact further down the ball, the ball will have a steeper initial trajectory with a lower velocity, but the rate of spin will increase.

A similar situation occurs when a striker hits a hard shot with the outside of his or her foot, slightly off center. The net force vector from the striker's foot passes through the center of the ball and determines the initial path of the ball. In this case, the net force vector has two components, one in the direction of the foot's movement and one perpendicular to the foot's movement. Experienced players can cause not only backspin and sidespin, but they can also cause spins in virtually any direction by controlling where they hit the ball and what the path of their foot is relative to a line through the center of the ball.

Once the kick has been taken and the spinning ball is flying forward, the fun really begins! Recall our earlier discussion about symmetry, airflow, boundary layers, and conservation of energy and momentum. All of these factors come into play as we consider what happens as a spinning ball flies through the air.

First, we will use the principle of analytical symmetry and let the ball be at rest with the air moving past it at a velocity equal to that of the ball when it left the player's foot. As the ball spins, the boundary layer—that very thin layer of air immediately adjacent to the ball's surface—spins along with it. On half of the ball, the air in the boundary layer is moving in the same direction as the surrounding airflow and therefore is moving faster. On the other half of the ball, the air is moving in an opposite direction to the surrounding airflow and therefore is moving more slowly. As we noted earlier, if the speed of the air is decreased, the pressure is increased and vice versa. Therefore, a pressure force pushes the ball in the direction in which it is spinning.

Now we add a new factor—viscosity. **Viscosity** is a characteristic of all fluids. Think about pouring water from a jar—it flows quickly.

Now think about pouring honey from a jar—it flows much more slowly. This "stickiness" is viscosity. Viscosity is caused by the tendency of the molecules in a fluid to stick to each other as they move. Since air is a fluid, it also has viscosity, although its viscosity is much less than that of honey, catsup, or even water. However, when we consider spinning soccer balls and the banana kick, the viscosity of air becomes important.

Consider the spinning ball in the diagram. At the top of the diagram where the spin is in the same direction as the surrounding airflow, the velocity of the air is highest and the pressure is lowest. At the bottom of the diagram where the spin is in the opposite direction to the surrounding airflow, the speed of the air near the ball is low and the pressure is highest. In addition, at some point near the bottom of the ball, the air in the boundary layer next to the ball is moving so slowly that the air molecules moving toward the ball have less of a tendency to stick together and simply bump into the ball and are deflected downward. This is called the point of flow separation. Since the molecules of air are bumping into the ball and being deflected downward, they are giving up some of their energy and momentum to the ball (think of a collision between a cue ball and another ball on a pool table). By Newton's laws and the principle of conservation of momentum, the ball must be "pushed" away from the airstream that is being deflected downward. Thus, it is pushed in the direction of the spin.

(1) At this point the air in the boundary layer next to the spinning ball is moving with the overall air flow past the ball, so this is the point of highest air velocity and lowest air pressure.

(2) At this point the air in the boundary layer next to the spinning ball is moving in the opposite direction to the overall air flow past the ball, so this is the point of lowest air velocity and highest air pressure. It is also the point of flow separation where the airstream is deflected downward.

(3) Since the pressure is highest on the bottom and lowest on the top, the ball is pushed in this direction.

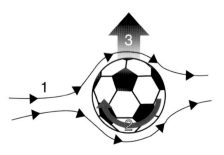

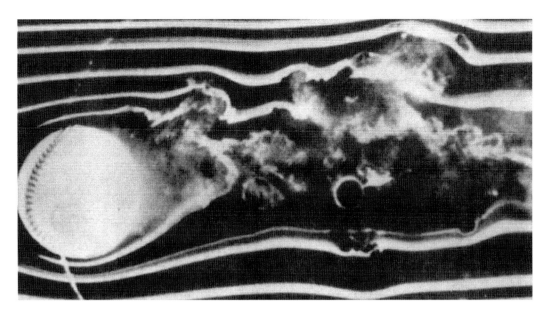
Flow around a counterclockwise-spinning baseball

The combination of the differential pressure and the flow separation leads to a force on the ball that is called the **Magnus effect** after the military ballistician, Heinrich Magnus.

Within the range of ball velocities and spin rates that are possible in soccer, the amount of lateral movement of the ball will increase as spin rate and velocity increase. Kicks by elite professional players have been clocked at up to 36 meters per second (80 mph), and kicks from 14-year-olds have been measured at well over 13 meters per second (30 mph). Lateral movement of as much as four meters (about 12 feet) has also been observed on kicks by elite professional players.

If a player hits the ball on one side or the other, he or she can induce sidespin. Using sidespin, the Magnus effect will cause a ball to bend around a wall (the "bend it like Beckham" phenomenon). In the case of a player who kicks the ball below the center of mass as described earlier, backspin is induced, causing a net upward force on the ball, increasing its range. Goalkeepers can extend the distance traveled by kicks by several meters because of the backspin. This is one reason that players kick at an angle less than 45°—the additional range provided by the Magnus effect is greater than that caused by kicking at a steeper angle.

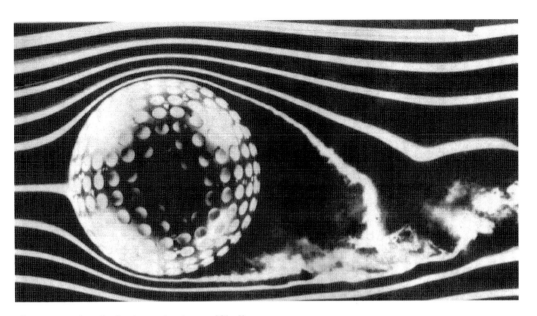

Flow around a clockwise-spinning golf ball

Obviously if there is wind, it will change the trajectory of the ball. However, the effect of wind can be handled by simply adding the effective speed of the ball to the wind speed to obtain a net flow of air past the ball. This can either enhance or detract from the Magnus effect, depending on the direction of the wind relative to the kick.

In baseball, pitchers can throw a pitch called a knuckle ball that has essentially no spin. Soccer also has its knuckle ball equivalent. Consider the case where the force vector of the foot runs exactly from the point of impact through the center of mass of the ball and where there is no net friction between the foot and the outer surface of the ball during the kick. In this case, no spin would be induced. When this happens, the boundary layer can be strongly affected by minor variations in the roughness of the surface of the ball. If one side has more stitching than the other, the boundary layer will tend to separate, pushing the ball away. If there is just a slight degree of rotation, the point of separation may move from side to side as the rougher side moves. This can cause the ball to appear to jump back and forth during flight. In practice, it is difficult to kick a ball with such a small amount of spin that one sees this "knuckle ball" effect, but if such a kick can be mastered, it can really challenge a goalkeeper!

Experiment—*The effect of kicking*

Purpose Examine the results of striking a ball at various locations.

Equipment required Soccer ball, stopwatch, tape measure

Experimental procedure Mark the equator (the location of maximum circumference) of a standard soccer ball. Make visible marks two and four inches down from the equator. Place the ball on the ground such that the marked equator is parallel to the ground. Kick the ball with the same force and foot movement (as nearly as possible) at the equator and each of the two marks. Repeat the kick three or four times at each mark.

Questions to be answered in this experiment
1. How far did the ball travel before it hit the ground for the first time?
2. How long was the ball airborne?
3. What was the approximate initial speed of the ball after it was kicked?
4. Which kicks gave the greatest spin?

Experiment—*Demonstration of the Magnus effect*

Purpose This experiment illustrates the Magnus effect.

Equipment required Two large Styrofoam coffee cups, Scotch tape or glue, two large rubber or elastic bands

Experimental procedure Use the tape or glue to stick the bottoms of the two cups together. Knot the rubber or elastic bands together. Hold the bands in the center of the cups and wrap them around two or three times. Hold the cup in one hand and the end of the rubber or elastic in your other hand. Pull back the cups and release them. With enough practice you should be able to make the flying cups loop in the air.

(See Appendix A, Experiment #4C on page 73 for further discussion.)

Tape or glue the cups together at the bottom

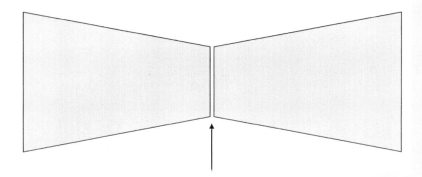

Bands wrapped around the cup, starting at the top and going around clockwise with the end coming toward you.

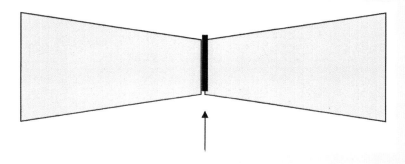

CHAPTER 5

Cracking the Whip

So far, we have talked about the history of the game, some basic scientific principles, and the behavior of the ball, both bouncing and moving through the air. Now we will look at two other aspects of soccer science—biomechanics and biochemistry. In other words, how do the various parts of your body move together to allow you to kick, head, or throw a ball? How do the concepts of vectors, kinetic energy, potential energy, and conservation of energy apply to the movement of the body on a soccer field? The answers to these questions come from the science of **biomechanics**.

Running, jumping, kicking, heading, and throwing all require energy to be used by the body. This energy is provided by the body's fuel and power generation systems. So, where does this energy come from? How does the food we eat, the water we drink, and the air we breathe turn into energy to make our muscles do what we ask them to during a game? Why are we so tired at the end of a game? The answers to these questions come from the science of **biochemistry**.

Have you ever played crack the whip on ice skates or Rollerblades? If so, you know that the people on the inside of the circle move relatively slowly while those near the outside fly around the circle much faster. The three ways of propelling a soccer ball—kicking, heading, and throwing—are a lot like cracking the whip.

Imagine a player preparing to kick a ball with his or her right foot. The player runs toward the ball at about a 45° angle to the direction that he or she wants the ball to go. Approaching the ball, the player plants his or her left foot beside it, pointed in the direction of the intended path of the ball. This transfers some of the player's kinetic energy to the ground as the sod is compressed. A combination of the kinetic energy of the run-up and contraction of the large hamstring muscle on the back of the player's leg "cocks" the right leg behind him or her. A pivoting move around the left hip, initiated by the hip flexor muscles, begins to move the player's right hip and right thigh, just like the skaters on the inside of the "whip."

If we could stop the action at this point in time, we would see the player's left foot planted to the left of the ball and his or her right leg cocked behind, right foot almost at waist level. The player is just beginning to pivot around his or her left hip in the horizontal plane (i.e., the plane parallel to the ground). Contraction of the hip flexor muscles is starting to pull the player's right thigh forward.

The kicking sequence. Lines show sequential leg position.

As the right thigh swings forward and down, using some of the kinetic energy of the run-up and also using contractions of the quadriceps muscles on the front of the player's thigh, the knee begins to move forward, lagging the hip but catching up quickly. As the right knee passes over the ball, the quadriceps muscles on the front of the thigh accelerate the lower leg forward, and the foot strikes the ball on the instep between the center of the foot and the ankle.

A lot of energy is involved in kicking a soccer ball. Studies have shown that about 15 percent of the energy of the swinging leg is transferred to the ball with the rest used to slow the leg down after contact. On a typical kick, the foot is in contact with the ball for about 10 milliseconds (0.01 seconds). The longer the contact, the greater the impulse of the foot on the ball. Similarly, the greater the distance that the force of the foot acts through, the more work done on the ball and the more energy transferred from the foot to the ball. Everyone has experienced how much energy is transferred to the ball during a kick—think about a time when you tried to kick a rolling ball really hard but completely missed. You may have fallen on your backside or completely spun yourself around, but for sure it hurt (and may have been a bit embarrassing as well)! The energy that would have caused the ball to fly off helped to put you on the ground!

The biomechanics of heading are a lot like kicking, except that they involve the torso, the neck, and the head instead of the hips, thighs, calves, and feet. The heading sequence also has a set of steps—a run-up or jump, a "cocking" of the body, a movement toward the ball, the striking of the ball, and a small follow-through. However, heading is not done with a stationary ball and does not generate as powerful a strike as does kicking.

There are three types of headers—one in which a player hits an airborne ball directly and with force; one in which the player strikes a lower ball by diving toward it; and the flick or glancing header in which the player strikes a moving ball just to change its trajectory.

During a "standard" header, the player runs toward the location where she intends to jump toward the ball. This builds kinetic energy. As the player reaches her jump-off point, the powerful gluteus and quadriceps muscles push her upward toward the ball. At the same time, she is bending backward at the waist, effectively "cocking" her upper body (i.e., torso, shoulders, and head). Holding her head firmly with her neck muscles, she swings her shoulders and torso forward, striking the ball on her forehead, right at the hairline. The contact

time for heading is a bit longer than that experienced during a kick (20–25 milliseconds for a header as compared with 10 milliseconds for a kick) because the head is moving more slowly than the foot, so the ball has more time to compress against the head and rebound.

Most of the kinetic energy of the run-up becomes potential energy as the player jumps into the air, although a small amount may be transferred to the ball. A significant amount of the kinetic energy of the whipping torso, neck, and head can be transferred to the ball. Elite professional players can add as much as 4.5 meters per second (10 mph) to the speed of a ball with a properly timed jumping or standing header.

In the case of a diving header, the main source of energy is the kinetic energy of the diving individual, since there is little or no whipping action from the torso or neck of the player. However, only a small fraction of the kinetic energy of the player is actually transferred to the ball; most of it is taken up in slowing the player when he or she hits the ground and skids after the dive.

Glancing or flicking headers are not designed to add energy to the ball but to redirect it so that it goes to a teammate or into the goal.

Another means of propelling the ball is the throw, either a throw-in from the touch line or a throw by a goalkeeper who has recovered the ball in the penalty area. There are three types of throw-ins—a simple toss from a standing position to a nearby player, a running throw that is designed to get the ball near the center of the field, and a so-called flip throw that is designed to get even more distance and velocity.

The biomechanics of a simple throw-in are quite obvious. A running throw is more complex, but has a lot in common with a header. In the case of the running throw, the player transfers some of the kinetic energy of the run toward the touch line into a crack-the-whip movement that originates with the player's torso and moves to his or her arms and then to the ball.

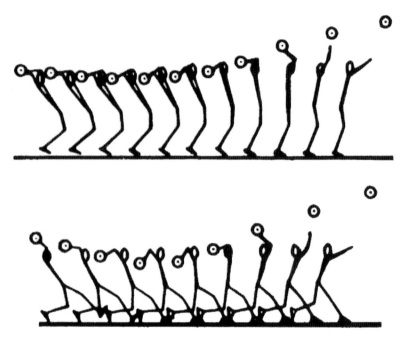

Throw-in sequence (standing—top, running—bottom)

The so-called flip throw, originally developed by a player at the Fountain Valley School in Colorado, involves a player running toward the touchline, performing a front handspring while holding the ball, and whipping the ball forward as he lands and completes the handspring. The crack-the-whip aspects are obvious—the initial pivot point is the ball on the ground, and as the player's legs move over his head and toward the ground, his body bends backward. When the feet land, the ball is just coming off the ground with the body in a backbend configuration, coiled like a spring from the ankles to the wrists. Almost all of the energy involved in the run-up, plus the energy represented by the cocked muscles in the legs, torso, and arms, is transferred to the ball. An expert flip thrower may actually just fall to the ground or even remain upright after the throw, suggesting that a maximum amount of energy has been transferred to the ball. As a result, the ball flies off with tremendous speed.

Studies have shown that a conventional throw-in can propel the ball at an initial speed of up to 18 meters per second (60 feet per second) for a range of about 30.5 meters (100 feet), whereas a flip throw can propel the ball at an initial speed of 23 meters per second (75 feet per second) for a range of nearly 46 meters (150 feet). When a flip throw is taken near the penalty area, it can be compared to a corner kick in terms of velocity and tactical advantage.

A throw by the goalkeeper involves cracking the whip as well. A goalkeeper throw is not like a baseball pitch, but resembles the overhand, straight-arm bowling motion of a cricket pitcher. The arm with the ball is fully extended behind the body and is pulled forward by a combined motion of muscles in the legs, torso, shoulders, and arms. Using this motion, and perhaps combining it with some run-up to the outer edge of the penalty area, accomplished goalkeepers can throw the ball past midfield.

Flip-throw sequence

Trapping is designed to absorb the energy of a ball. In this case, the player wants to slow or stop a ball that is coming to him either in the air or on the ground. Two common types of traps are chest traps and foot or leg traps. Recall the difference in rebound or bounce behavior of a ball when it strikes a hard surface like a goalpost, versus when it strikes the field. Obviously, the harder the surface, the less energy is absorbed by the surface, leading to a higher coefficient of restitution as discussed in chapter 2. If less energy is absorbed, more is retained as kinetic energy of the ball as it rebounds from the surface. Now extend this principle to a ball striking parts of the body. If you want to stop or slow down a ball, do you want it to hit a hard part of the body like your shin or a softer part of your body like your calf or thigh? Do you want to force your chest out, putting your ribs and tensed pectoral muscles right against the ball, or do you want to collapse your chest as the ball hits you, allowing the energy to be absorbed over a longer distance and time, thus providing more negative impulse and reducing the work?

In the case of a chest trap, you are taught to collapse your arms inward, causing your upper body to move backward just as the ball hits you. This has the effect of softening the surface that the ball is

Thigh trap   Foot trap

striking, meaning that the energy is absorbed by the body and the ball slows down, hopefully dropping right at your feet. In terms of the physics, you have both provided a smaller coefficient of restitution by collapsing your body as the ball hits it, and you have increased the time of contact, thereby reducing the impulse.[*] For these reasons, the rebound velocity is much less than the velocity before the ball struck your chest.

Another means of stopping the ball is a catch by the goalkeeper. Have you ever wondered why goalkeepers frequently decide to punch or parry the ball rather than catching it? Think about how hard some shots can be—up to 36 meters per second (80 mph) in professional ranks! The energy involved in stopping a shot like this would be totally absorbed by the goalkeeper and, by conservation of energy, might actually be sufficient to push him or her backward into the goal. However, if the goalkeeper provides a rigid surface like a hand or fist (that is, a high coefficient of restitution), he or she can

---

* Recall that impulse equals force times time. Although in this case we are increasing the time, we are decreasing the force, thus resulting in a negative component of impulse.

redirect the ball to a teammate downfield with most of its kinetic energy intact.

Now let's examine how the food we eat and the air we breathe are converted into running, jumping, and kicking. First the food, water, and air must somehow be made into fuels that can be used by the body. Then the fuel must be used by the cells to generate energy. Then the energy must be used by the muscles to make them contract and relax. And, of course, all of this must be controlled by the brain as the players decide whether to sprint to the ball or kick it into the goal.

The process starts with digestion, which turns food into material to build tissues or into fuel to allow cells to function. In order for the body's energy-generation mechanism to work, there must be a supply of basic materials—the oxygen, sugar, water, nitrogen, calcium, phosphorus, and other trace elements that are needed for various chemical reactions. Oxygen comes in through the lungs, so increased demand for oxygen results in more rapid breathing. The

sugars and other materials come from food and are stored in the body for use when needed.

Your body converts some of the foods you eat into a chemical called **glycogen** which is stored in the liver and in muscle tissue. Glycogen can be simply thought of as a storage molecule—a long string of sugar molecules storing energy in the form of carbohydrates. When you exercise, the body can release the sugar stored as glycogen very quickly. Although you can get a quick sugar fix with a candy bar, the only way to provide sustained energy is with more complex carbohydrates. These are easily converted into sugars by the body, so they provide a ready source of sugar to make glycogen.

Water is required for at least three reasons. First of all, many bodily functions like digestion require water. Secondly, some of the chemical reactions needed to produce energy require the presence of water. Thirdly, and most importantly, perspiration helps us maintain body temperature when we are exercising. The sweat produced by the body evaporates from the skin, and this evaporation

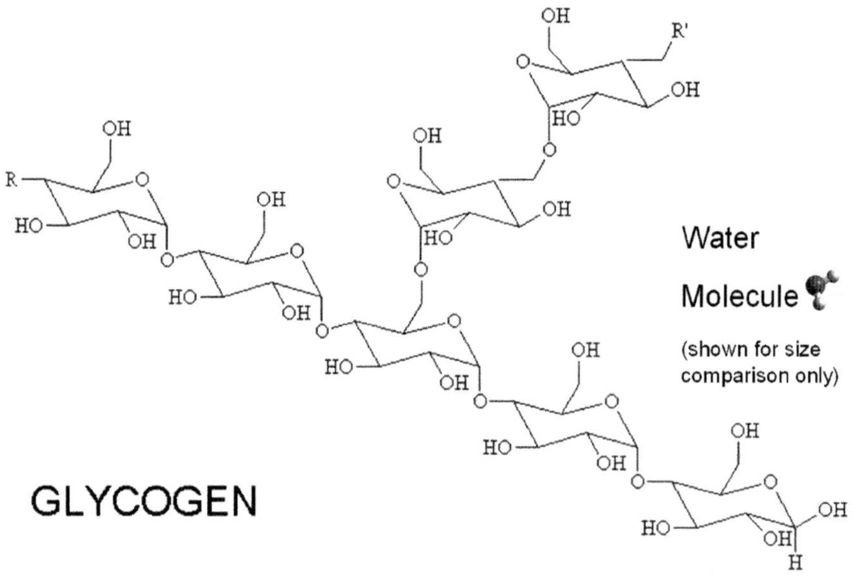

Water Molecule

(shown for size comparison only)

GLYCOGEN

cools the body. This process is quite efficient when the humidity is low, but it can require much more perspiration when the temperature and humidity are both high. Studies have shown that players can lose up to 3 percent of their body weight in perspiration in an intense match under hot, humid conditions. For these reasons, soccer players must drink plenty of water, both before and during a match.

Digestion provides the fuel for our cells to work, and the energy that we develop in our bodies ultimately comes from another energy-storing chemical called **adenosine triphosphate** (ATP). When the muscles receive a nerve signal to contract, the ATP in the muscle cells is converted to **adenosine diphosphate** (ADP), releasing the energy stored in the ATP.

ATP is produced in one of several ways: the ADP produced in muscle cells can be converted back to ATP using a chemical called **phosphocreatine**; the glycogen stored in muscle cells and in the liver and the glucose in the bloodstream can be converted to ATP using a process called **glycolysis**; and both glycogen and fats (in the form of free fatty acids) can be combined with oxygen in mitochondrial cells to form ATP in a process call **aerobic metabolism**.

The release of energy from the ATP-ADP reaction allows the thick muscle filaments (called **myosin**) to slide over the thin muscle filaments (called **actin**), causing the muscle to contract. When the contraction is complete, the body signals the muscles to relax by increasing the level of calcium in the tissue. This causes the contraction to stop and allows the thick filaments to slide back in the other direction, causing the muscle to relax. When you combine thousands of muscle filaments all contracting or relaxing together under the direction of a conscious thought like "run," "jump," or "kick," you get the actual motion desired.

If you have eaten properly before a match and have drunk enough water, you will be well hydrated and will have the necessary

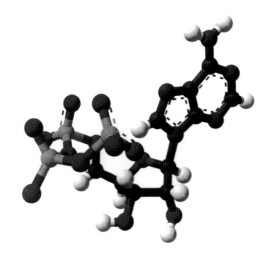

ATP (*Black*: Carbon, *White*: Hydrogen, *Red*: Oxygen, *Blue*: Nitrogen, *Orange*: Phosphorus)

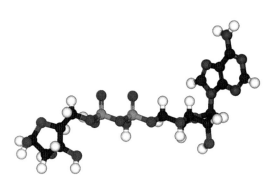

ADP

glycogen in muscle cells to produce the required amount of ATP throughout the match. However, if you have not eaten enough or if the game is particularly long or physically challenging, your body experiences fatigue. This is a complicated phenomenon, but there are two basic causes. In the first case, your body is using more ATP in the myosin-actin operations than it is producing from the ATP-ADP reaction. This causes your muscles to perform less efficiently. The other situation results from a low level of glucose in the bloodstream. This condition, called **hypoglycemia**, not only affects the amount of ATP in muscle tissue but also causes overall fatigue, including "central fatigue," which can affect focus, motivation, and overall performance. Both of these phenomena are a direct result of insufficient glycogen in the muscle tissue.

Computing the amount of energy actually used by an individual while he or she is running depends on several variables, including the individual's weight, personal metabolism, running efficiency (a function of stride length, leg strength, and other factors), and how well he or she can get rid of the waste heat generated during the chemical reactions described above.

Laboratory studies suggest that a 68-kilogram (150-pound) player running at 2.7 meters per second (6 mph) uses about 700 food calories per hour. If the player weighs more (or less) than 68 kilograms,

The definition of a **calorie** is the amount of energy required to raise 1 gram of water 1° Celsius. This is a very small amount of energy, and the amount of energy used by the body while exercising (and taken in in the food we consume) is very large. Therefore, when we speak of **food calories** (those consumed by eating or burned while exercising), we are really talking about kilocalories (1,000 calories). A **kilocalorie** or food calorie is the amount of energy required to raise 1 kilogram of water 1° Celsius.

A person running a 4-minute mile is moving at about 6.7 m/s (15 mph), and the world's fastest sprinters run at about 8.9 m/s (20 mph), although for less than a minute or so in races like the 400-meter dash.

you should add (or subtract) 60 food calories per hour for each 4.5-kilogram (10-pound) deviation. For example, a 54.5-kilogram (120-pound) player would burn 520 food calories if he or she ran at a 2.7 meters per second (6 mph) pace for one hour. Similarly, the speed at which you run makes a difference. For each 0.45 meters per second (1 mph) above a 2.7 meters per second (6 mph) pace, a person will burn about 100 additional food calories per hour.

A study of professional soccer players suggests that they move, on average, between 8,800 and 10,000 meters per game with about 67 percent at a walk or a jog and only about 10 percent at a full-out sprint. Goalkeepers move less than average, and midfielders more than average. Youth players run or jog about 48 percent of the time and only spend 1 percent of their time in full-out sprints. This means that in a 90-minute game, the average youth player will burn between 500 and 1,000 food calories (about the equivalent of a Quarter Pounder with Cheese and a small order of fries!).

CHAPTER 6

Putting It All Together

In order to put everything into a real-life situation, let's return to the scenario from the beginning of chapter 1. This time, instead of beginning with the free kick, let's start with a goalkeeper save at the other end of the field. As the play unfolds, we will describe the action in terms of biochemistry, biomechanics, and physics, not just tactics and strategy.[*]

The keeper saves a shot on goal and jogs to the edge of the penalty area, looking upfield for an open player. A midfielder is unguarded near the center circle. The keeper steps back to avoid stepping out of the penalty area then takes two quick steps, using his right hamstring muscle to cock his right leg. As he drops the ball, his quadriceps and hip flexors, fueled by ATP-ADP reactions in thousands of muscle cells, accelerate his lower leg and foot. As his thigh passes through the vertical plane, the lower leg snaps forward, allowing the bony top of his foot to contact the ball just before it hits the ground. The keeper's foot is in contact with the ball for about 10 milliseconds, compressing it to about half its diameter. As it rebounds from the compression, the ball is driven upward and forward at about 22.3 meters per second at a 45° angle. After the ball has left his foot,

---

[*] The actual calculations for this scenario (in both metric and English units) can be found in Appendix B.

the hamstring muscles absorb the remaining kinetic energy of the leg, slowing it down.

The ball rises to a height of about 12.8 meters in about 1.6 seconds. As it drops toward the ground, it is intercepted by the midfielder who has jumped to hit the ball with his head at a height of about 2.4 meters above the ground. This means that the ball falls for about 1.5 seconds. In the 3.1 seconds after the kick, it travels a horizontal distance of about 36.6 meters to reach the midfielder near the center circle.

The midfielder runs a few steps and jumps toward the ball, cocking his head. His head strikes at the hairline on his forehead and the snap of his torso and neck muscles cause about 2.7 meters per second to be added to the ball's horizontal component of velocity as he sends

it toward the left wing, who is sprinting down the field. The head strike has a restitution factor of about 0.6 because of the absorption of energy in the neck and torso muscles, so the ball leaves the midfielder's head at about 13.8 meters per second at an angle of 39°.

The ball flies toward the wing, who is sprinting at a speed of about 7.6 meters per second from a point in his own half of the field. The ball reaches a height of about 3.7 meters and travels about 21 meters in 2.0 seconds to a point just inside the touch line, where the wing allows the ball to hit the inside of his right calf in such a way that the energy is almost completely absorbed. Continuing his run, the wing keeps the ball on the ground a few meters in front of him to a point just even with the outer edge of the opponent's goal area, where he is confronted by a defender. He covers the 45 meters in about 5.9 seconds and burns about 3 food calories in the effort.

Without breaking stride, the wing plants his right foot and allows most of the kinetic energy of his sprint to add to the energy provided by his quadriceps to move his left leg around. He strikes the ball, driving it toward the top of the penalty area at about 26.7 meters per second at an angle of about 45°.

The ball travels about 45.7 meters over the ground in about 3.9 seconds, reaching a striker on the far side of the penalty area above the arc, about 25 meters from the goal. Just as it does, however, a defender fouls the striker, the whistle blows, and the referee signals for a direct free kick for the attacking team.

The defenders quickly form a wall 10 meters from the ball and, at the direction of the goalkeeper, move to their right to cover the far post while he covers the near post. The striker who is taking the kick surveys the situation from a position about five meters from the ball. Making his decision, he accelerates forward, plants his right foot, and strikes the ball hard just to the left of its center. The ball leaves his left foot at about 26.7 meters per second with an initial trajectory just to the left of the wall, spinning clockwise at a rate of about 800 revolutions per minute. The goalkeeper sees the ball headed around the wall and feels momentarily relieved—this kick will go well beyond the goal and should be no problem.

The ball passes the wall after about 0.4 seconds. The clockwise spin of the ball that the kicker introduced by his off-center strike causes the speed of the air in the part of the boundary layer farthest from the wall to be about 17.3 meters per second, more than enough to allow the flow to separate from the ball and be deflected in a direction away from the last player in the wall.

On the side of the ball closest to the wall, where the boundary layer flow caused by the spin adds to the speed of the surrounding air, the air is moving at 36.3 meters per second. This combination of the pressure difference due to the velocity difference and the momentum transfer caused by the flow separation (the Magnus effect) pushes the ball to its right, causing it to curve about 2.6 meters to the right around the last player in the wall. In addition, aerodynamic drag causes it to slow down. About 0.6 seconds later it flies into the goal, just inside the post.

This entire sequence, from a goalkeeper save to a goal scored on the opposite end of the field, has taken about 60 seconds, with the largest amount of time being the 30 seconds taken to set the wall and take the free kick. Since a soccer game is 90 minutes long, there can be 90 of these magnificent sequences of physics, biomechanics, biochemistry, and aerodynamics in every game. Certainly, this is a beautiful game!

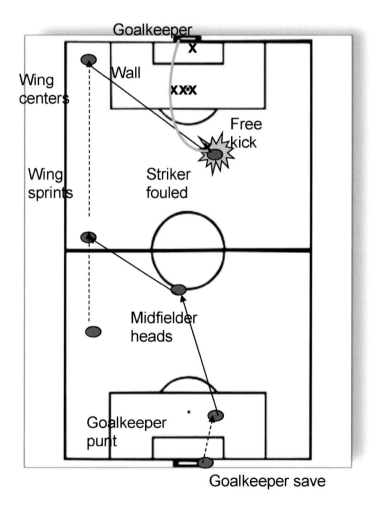

Goalkeeper

Wall

X

xxx

Wing
centers

Free
kick

Wing
sprints

Striker
fouled

Midfielder
heads

Goalkeeper
punt

Goalkeeper save

Appendix A

Verifying What You Have Learned

We have taken a broad look at the history and science of the game of soccer. For those of you who only want the basic material, you have probably read far enough! For those who would like more information, please refer to the reference material in the bibliography. In this appendix we provide some experiments that demonstrate what we have discussed. In each case, the basic experiment is described, together with some suggestions for how you might conduct it and a more detailed explanation. We suggest that you try the experiment yourself before reading the summary material. Good luck!

Experiment #1—Follow the bouncing ball

PURPOSE

This experiment illustrates the behavior of a bouncing ball. It demonstrates various aspects of elasticity and restitution.

EQUIPMENT REQUIRED

Soccer ball, baseball, pool or croquet ball, inflation pump and pressure gauge, various surfaces (concrete, grass field, etc.), yardstick

EXPERIMENTAL PROCEDURE

Drop each of the balls from a fixed height onto various surfaces and measure the height of the rebound. Vary the height from which the ball is dropped, the pressure in the ball, and the surfaces on which the ball is dropped. Make several measurements at each drop height, pressure, and surface. Graph the rebound height (y axis) as a function of drop height (x axis).

QUESTIONS TO BE ANSWERED IN THIS EXPERIMENT

Which ball bounces highest? Why?

How does the height of bounce depend on the surface?

How does the height of bounce depend on the pressure of the ball?

Can you derive an expression for the coefficient of restitution that relates the ratio of the drop and rebound heights to the drop and rebound velocities?

DISCUSSION

The coefficient of restitution is the ratio of the velocity of the ball after rebound to its velocity before striking the surface. However, it is much easier to measure the height of rebound than the rebound velocity.

In order to derive the relationship between the height and the velocity, we use conservation of energy. When the ball is being held at some height, h, just before being dropped, it has no kinetic energy (that is energy of motion), but it has potential energy that is given by the formula

$E_p = mgh$, where E_p is the potential energy, m is the mass of the ball, g is the acceleration of gravity, and h is the height.

If we ignore air friction (a good assumption for small heights and relatively low velocities), the energy of motion of the ball when it is moving is

$E_k = \frac{1}{2} mv^2$, where E_k is the kinetic energy, m is the mass of the ball, and v is the velocity of the ball.

We know that energy must be conserved, so $E_p = E_k$ when the ball strikes the ground. In other words

$mgh = ½ mv^2$.

Therefore, the relationship between h and v is given by

$h = v^2/2g$ or $v = \sqrt{(2gh)}$.

Now, we can compute the coefficient of restitution, e, which is the ratio of the velocity after rebound to the velocity before rebound:

$e = v_2/v_1 = \sqrt{(h_r/h_d)}$, where v_2 is the velocity after rebound, v_1 is the velocity before striking the surface, h_r is the height of the rebound, and h_d is the height of the drop.

Experiment #2—The effect of drag

PURPOSE

These three experiments illustrate the effect of drag on a soccer ball.

Experiment #2A illustrates the drag felt by the ball when it passes through a medium like the atmosphere. We will use much denser and more viscous materials like oil, water, and shampoo to show the effects of drag more dramatically.

Experiment #2B illustrates the drag felt by the ball when it is rolling.

Experiment #2C quantifies the effect of drag using a high-speed camera.

EQUIPMENT REQUIRED

Experiment #2A: Glass bottle or test tube, marble, stopwatch, various liquids (like water, motor oil, shampoo, or syrup)

Experiment #2B: Marble, stopwatch, wooden ramp with edges (five to six feet long would be ideal), stopwatch, various materials (grass, carpet, concrete, dirt)

Experiment #2C: Soccer ball, high-speed or framing camera, stopwatch

EXPERIMENTAL PROCEDURE

Experiment #2A (fluid dynamic drag)—Fill the glass container with one of the fluids. Release the marble into the container and measure the time it takes to fall to the bottom. One of these experiments should be done with an empty container (i.e., a container filled only with air). Graph the time required for the marble to fall to the bottom (*y* axis) as a function of the density of the material (*x* axis). You will have to ask your science teacher for logarithmic graph paper for this experiment.

Experiment #2B (rolling/surface drag)—Roll the ball down the ramp onto the various surfaces. Measure the time that the ball takes to move the same distance on each of the surfaces. Try performing this with the surfaces both dry and wet.

Experiment #2C (quantifying drag)—Set up a framing camera perpendicular to the path of a kick and far enough from the path of the ball so that the initial kick will appear in at least two frames. Determine the speed of the ball immediately after it is kicked by measuring the distance traveled and dividing by the framing rate. Compare this value with the actual time for the kick to travel its full distance.

QUESTIONS TO BE ANSWERED IN THIS EXPERIMENT

What is the effect of drag on the movement of an object?

Does the effect of drag change with the material that the ball is passing through?

The marble dropping through various fluids illustrates fluid dynamic drag. The rolling experiment illustrates friction and skidding. The framing camera experiment quantifies the effect of drag on velocity.

Experiment #3—Conservation of energy in fluid flow

PURPOSE

This experiment demonstrates conservation of energy in a fluid like air.

EQUIPMENT REQUIRED

A large fan, a beach ball, a child's pinwheel and/or a handheld anemometer (an instrument that measures wind speed)

EXPERIMENTAL PROCEDURE

Place the fan horizontally across two chairs or two sawhorses so that the airflow is vertical rather than horizontal. Turn the fan on and place the inflated beach ball in the airstream. Try to move the ball out of the airstream by gently tapping it. If the fan is equipped with a variable speed control, vary the speed and graph the height of the ball as a function of the speed of the airflow.

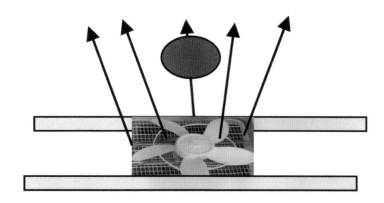

Why does the beach ball remain in the airstream?

What forces are acting on the beach ball?

What is the effect of varying the speed of the airflow?

DISCUSSION

The forces acting on the ball are (1) gravity and (2) the force of the air molecules impinging on the ball and pushing it upward. The second force can be described as the drag force, since it is "dragging" the ball up against the force of gravity, which is trying to make it fall to the ground. The drag force is given by

$F = \frac{1}{2} C_d \rho A v^2$, where C_d is the drag coefficient, ρ is the density of the air, A is the cross-sectional area of the ball ($A = \pi r^2$), and v is the air velocity.

The second force is the force of gravity. Since force equals mass times acceleration, the force of gravity can be expressed as

$F = mg$, where m is the mass of the ball and g is the acceleration of gravity.

Since the forces must be equal and acting in opposite directions if the ball is stationary in the airstream,

$$mg = \frac{1}{2} C_d \rho A v^2$$

so the velocity necessary to keep the ball suspended in the airstream is given by

$$v = \sqrt{(2\ mg)/(C_d \rho A)}.$$

Now, consider the air column above the fan. Right at the fan the velocity of the air is highest and at some point above the fan you can no longer feel it moving. This is because the air pushed by the fan spreads out in a conical manner, eventually losing all of its velocity. You can see this using a child's pinwheel. Move the pinwheel back and forth across the flow stream at various heights and notice how the speed increases toward the center and decreases toward the edges. As you move further up from the fan, you will see that the overall speed slows down and the moving airstream is farther and farther from the centerline of the fan. You can get a measurement of the velocity of the airstream at various locations using a handheld anemometer, a device that measures wind speed.

Now, if you increase the speed of the fan, the velocity necessary to hold the ball suspended will occur at a greater distance from the fan, so the ball will move up. If you reduce the speed, the ball will move down.

Recall our discussion of the fact that as a fluid moves around an object it accelerates so that the streamlines continue to be aligned. As the fluid accelerates, the kinetic energy of the air increases. However, by conservation of energy, the resting or pressure energy of the airstream must decrease when the kinetic energy of the airstream increases. Therefore, the energy of the air just outside the conical flow pattern caused by the fan has essentially zero kinetic energy and a high pressure while the air in the high-velocity airstream has a high kinetic energy and a lower pressure. As you move toward the edges of the airstream where the speed drops off, the pressure increases. Thus, if the ball moves toward the edge of the airstream, there is a pressure force pushing it back toward the centerline where the pressure is lower. This force is given by

$F = C_d (\frac{1}{2}\rho\, A)\, (v_e{}^2 - v_i{}^2)$, where v_e is the speed at the location nearer the edge of the airstream and v_i is the speed nearer the inside of the airstream.

Notice that the force is a negative one because v_i is greater than v_e, indicating that it is directed inward rather than outward.

Experiment #4—Demonstrating boundary layer phenomena and the Magnus effect

PURPOSE

These experiments will permit you to see boundary layers and observe the effects of factors like air speed, surface roughness, and spin on boundary layer phenomena.

EQUIPMENT REQUIRED

A large fan (in this case, a fan whose speed can be varied is desirable), a ball (this experiment will be easier if a tennis ball is used), a smoke-generating device, a variable-speed electric drill

EXPERIMENTAL PROCEDURE

Experiment #4A (visualizing boundary layers and boundary layer separation)—Set the fan so that it blows horizontally. Mount the soccer ball in front of the fan (I suggest attaching suction cups to opposite sides of the ball and attaching rods to a solid foundation that can be moved back and forth). Introduce the smoke into the airstream and observe the behavior as the streamlines form around the ball. Change the velocity of the airstream (either by moving it nearer or farther from the fan and/or by changing the speed of the fan) and notice any differences in the location and behavior of the streamlines around the ball. Note that as you move the ball closer to the fan, it will be harder to see the flow stream because of the turbulence introduced by the fan itself.

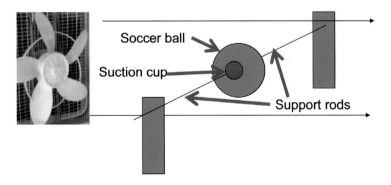

Experiment #4B (the effect of spin on boundary layer behavior: the Magnus effect)—With the fan set up to blow horizontally as in Experiment #4A, mount the ball with the suction cups and rods aligned horizontally, and arrange the rods so that the ball can rotate in the horizontal plane. Make one rod longer than the other such that it can be attached to a variable-speed drill. With the fan going and the smoke in the airstream, vary the speed of rotation of the ball. Observe the boundary layer behavior as the rotational speed of the ball changes.

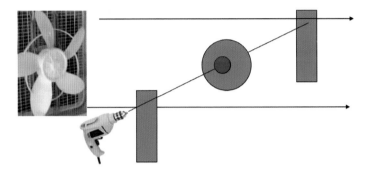

Experiment #4C (another demonstration of the Magnus effect)

EQUIPMENT REQUIRED

Two large Styrofoam coffee cups, Scotch tape or glue, two large rubber or elastic bands

EXPERIMENTAL PROCEDURE

Use the tape or glue to stick the bottoms of the two cups together.

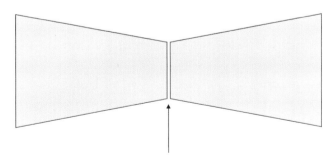

Tape or glue the cups together at the bottom

Knot the rubber or elastic bands together. Hold the bands in the center of the cups and wrap them around two or three times.

Bands wrapped around the cup, starting at the top and going around clockwise with the end coming toward you.

Hold the cup in one hand and the end of the rubber or elastic in your other hand. Pull back the cups and release them. With enough practice you should be able to make the flying cups loop in the air.

DISCUSSION

Experiment #4A—No further discussion.

Experiment #4B—No further discussion.

Experiment #4C—When the cups are released, the bands unwind, and the cups are forced to spin. If the bands are wound correctly the cups will be given backspin; the bottom of the cups move forward while the top is moving backward. Because of the rough surface of the cups, air is trapped in the boundary layer near the surface and moves with the cups as they spin.

The top of the cups has air moving from front to back as they spin, and the cups also have air flowing over them from front to back because they are flying through the air. The bottom of the cups also have air moving from the front to the back because they are flying through the air, but the bottom also has air moving back to the front because of the direction of the spinning cups. Therefore, the cups are sitting in air that is moving very differently at different parts: there is fast-moving air at the top while the air is close to being stationary at the bottom.

Because of the slower-moving air at the bottom, the flow separates and is deflected downward, transferring upward momentum to the cups. In addition, faster air has a lower pressure, so the cups have low pressure above them and higher pressure underneath. The net result is that the cups are forced upward.

Experiment #5—The effect of kicking

PURPOSE

Examine the results of striking a ball at various locations.

EQUIPMENT REQUIRED

Soccer ball, stopwatch, tape measure

EXPERIMENTAL PROCEDURE

Mark the equator (the location of maximum circumference) of a standard soccer ball. Make visible marks two and four inches down from the equator. Place the ball on the ground such that the marked equator is parallel to the ground. Kick the ball with the same force and foot movement (as nearly as possible) at the equator and each of the two marks. Repeat the kick three or four times at each mark.

QUESTIONS TO BE ANSWERED IN THIS EXPERIMENT

How far did the ball travel before it hit the ground for the first time?

How long was the ball airborne?

What was the approximate initial speed of the ball after it was kicked?

Which kicks gave the greatest spin?

DISCUSSION

How fast does the kicked ball travel? The average horizontal speed of the ball in flight can be determined by measuring the distance traveled and dividing it by the time the ball is airborne:

$v_h = d/t$, where v_h is the horizontal component of velocity, d is the horizontal distance from the point of the kick to the point of first impact, and t is the time required to travel that distance.

The horizontal speed is approximately equal to the average overall speed if the trajectory of the ball is approximately parallel with the ground. If drag is ignored, the average speed can be equated to the initial speed.

However, if the ball has significant vertical movement, we must consider both the vertical and horizontal components of the velocity vector. We can approximate the vertical velocity of the ball by first noting that the ball moves vertically only to the point where its kinetic energy (the energy of motion) in the vertical direction goes to zero. At this point, by conservation of energy, the kinetic energy associated with the vertical component of the initial velocity must equal the potential energy of the ball at the maximum height of the ball,

$\frac{1}{2}\,mv_v^2 = mgh$, where m is the mass of the soccer ball, v_v is the vertical component of the velocity, g is the acceleration of gravity, and h is the maximum height reached by the ball in flight.

When this is simplified and solved for v_v, we get

$$v_v = \sqrt{2gh}.$$

This would be fine except that it may be difficult to measure the maximum height of the ball. It is much easier to tell *when* the ball reaches its maximum height. If you can measure the time that it takes to reach its maximum height (this may take some practice, but you should be able to get close after trying it a few times!), you can use the expression

$v = at$, where v is the vertical velocity, a is the acceleration (in this case the acceleration of gravity, or g), and t is the time required for the ball to reach its maximum height. (Note that if you ignore aerodynamic drag and spin effects, the time to reach the maximum height is equal to the time required to fall from the maximum height or half the time from the point of the kick to the point of impact.)

We can now combine the two velocity components using the Pythagorean theorem for right triangles:

$$v = \sqrt{[(v_h)^2 + (v_v)^2]}.$$

Substituting the values that we computed

$$v = \sqrt{[(d/t)^2 + (gt/2)^2]}.$$

This gives you an approximation for the velocity of the ball when it leaves the kicker's foot. Compare the values that you get for the three marked locations on the ball. Can you approximate the 30 miles-per-hour value?

Of course, the reality is that the ball undergoes aerodynamic drag as it flies through the air, so both the horizontal and vertical components of the initial velocity that you computed are less than the actual components of the initial velocity. In addition, the effect of backspin will cause an upward force on the ball which will keep the ball in the air longer.

Appendix B

Calculations for Chapter 6

Initial Trajectory of Goalkeeper Punt

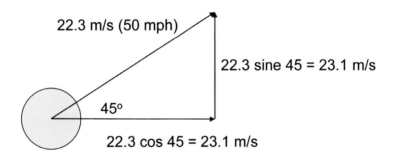

Maximum Height of Goalkeeper Punt (ignoring drag or Magnus effect)

Use the vertical component of velocity = v_v = 23.1 m/s (meters per second). Balance kinetic energy and potential energy

> *Equation 1*: ½ mv_v^2 = *mgh*, where *m* is the mass of the ball, *g* is the acceleration of gravity (9.8 m/s²), and *h* is the height of the ball.

At the point of maximum height, all of the kinetic energy has been converted to potential energy, so

> *Equation 2*: $h = v_v^2/2g$.

Therefore, the maximum height is 12.8 m (42 ft).

Time of Flight of the Ball
(ignoring drag and Magnus effect)

It can be shown that the time required for an object to travel a certain distance is proportional to the acceleration and time. The specific relationship is

Equation 3: d = ½ a t^2, where d is distance, a is acceleration, and t is time.

In this case a is the acceleration of gravity which is slowing the ball, so

Equation 4: $t = \sqrt{(2d)/g}$.

If the ball falls 10.4 m (34 ft) from its peak height to the midfielder's head, its vertical component of velocity when it strikes his head is given by equation 2, $v = \sqrt{2gh}$. In this case, v = 14.3 m/s (47 ft/s [feet per second]).

Substituting, we find that it takes 1.6 seconds for the ball to rise to its maximum height. This equation can also be used to determine how long it takes for the ball to return to ground level. However, by jumping, the midfielder has intercepted it about 2.4 m (8 ft) above the ground, so we use 12.8 m minus 2.4 m or 10.4 m (34 ft) as the distance in equation 4, and we find that it takes 1.5 seconds for the ball to reach the point at which it is headed by the midfielder. Summing these two, we find that the goalkeeper punt travels for 3.1 seconds.

Horizontal Distance Traveled by Goalkeeper Punt
(ignoring drag and Magnus effect)

We know the horizontal component of velocity and the time that the ball is in flight, so multiplying these values, we find that without considering drag, the ball would travel 49 m (161 ft) over the ground.

Incorporating Drag into Goalkeeper Punt Calculations

Dr. John Wesson's book *The Science of Soccer* provides a set of computer-derived graphs that incorporate the effect of drag on balls propelled through the air. Using these graphs, the actual distance traveled by the punt over the ground is about 36.6 m (120 ft) and the horizontal velocity at the end of the flight is about 16.5 m/s (38 mph).

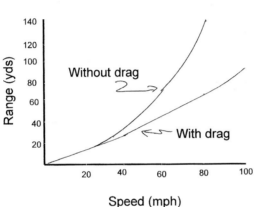

Range for kicks at 45° with and without drag

Midfielder Heading Calculation

When the ball is headed by the midfielder, it is traveling at 16.5 m/s (54 ft/s) with a vertical component of velocity of 14.4 m/s (47 ft/s). Thus the angle of impact is about 61° and the horizontal component of velocity is about 7.9 m/s (26 ft/s). The player adds a horizontal component of 2.7 m/s (9 ft/s) with his or her head, but if there is a coefficient of restitution of 0.6 for the vertical component, the new vector diagram becomes

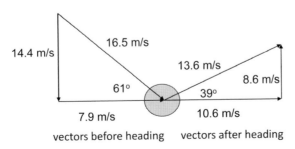

vectors before heading   vectors after heading

Maximum Height after Header

Using equation 2 with the new vertical velocity of 8.6 m/s (28.2 ft/s), we get a maximum height of 3.7 m (12 ft). However, the header began 2.4 m (8 ft) above the ground, so the maximum height of the ball is 6.1 m (20 ft).

Time of Flight of Headed Ball

In this case, we use 3.7 m (12 ft) for the rising portion of the trajectory and 6.1 m (20 ft) for the falling portion. Using equation 4 for the two distances, we get a rise time of 0.9 seconds and a fall time of 1.1 seconds. Summing, we find that the ball is in the air for 2.0 seconds between the midfielder's head and the winger's trap. In this period of time, the ball travels approximately 21.3 m (70 ft) over the ground (drag has been ignored in this calculation).

Food Calories Burned by Winger as He Sprints to the Ball

If the winger weighs approximately 68 kg (150 lb) and is running at 7.6 m/s (25 ft/s), he has a baseline (6 mph) food calorie consumption of 700 food calories per hour. The additional speed adds 1,100 food calories per hour for a total of 1,800 food calories per hour during the sprint.

If his sprint takes him from the halfway line to a point even with the outer edge of the goal area, and if the field is 100 m long, he must sprint for 50 m minus 5 m, or 45 m (148 ft). If he is moving at 7.6 m/s (25 ft/s), he covers the distance in about 5.9 seconds, and this sprint burns 3 food calories.

Winger's Kick

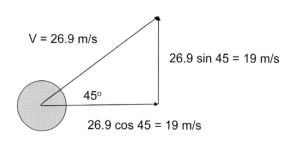

Height of Kick

Using equation 2, with the vertical velocity value of 19 m/s (62 ft/s), the ball rises to a height of 18.2 m (60 ft).

Time of Flight of the Ball

In this case the ball essentially moves from ground level to ground level, so we can simply use equation 4 and double the value. This gives a flight time of 3.9 seconds.

Distance Traveled

Multiplying the horizontal component of velocity by the time gives a distance traveled of 73.8 m (242 ft) without considering drag. When drag is included, the range is reduced to 45.7 m (150 ft).

Free Kick (Time and Curvature)

The free kick is moving at 26.7 m/s (88 ft/s), so the 10 meters to the wall is covered in 10 meters/26.7 m/s = .37 seconds. The remaining distance of about 15 meters to the goal is covered in 15/26.7 = .56 seconds (ignoring drag).

    The velocity at the surface of the spinning ball can be computed by noting that, in accordance with the rules of the game, the ball has a circumference of between 27 and 28 inches (68.6 and 71.1 cm). If we assume that the circumference is 68.6 cm and that the ball is spinning at 800 revolutions per minute, the tangential velocity at the outer surface of the ball is 9.4 m/s (31 ft/s). If the ball is moving forward at 26.7 m/s (88 ft/s), the air moving in the same direction as the spin (i.e., the air farthest from the wall) is moving at 36.1 m/s (119 ft/s) and the air moving in the opposite direction (i.e., the air closest to the wall) is moving at 17.3 m/s (57 ft/s).

    Equation 6 below is an approximation, derived by Wesson, for the curvature of a spinning kick that relates the curvature (D in feet) to the distance from the goal (L in feet), the spin rate (f in revolutions per second) and the velocity of the kick (V in feet per second [ft/s]) (see *Science of Soccer*, pages 167–69):

Equation 6: $D = (L^2/100)\,(f/V)$.

In this case, L is 75 ft (22.9 m), f is 13.3 revolutions per second (800 revolutions per minute), and V is 88 ft/s (26.7 m/s), so D is 8.5 ft (2.6 m).

Glossary

Acceleration: a measure of how fast an object (like a soccer ball) is either speeding up or slowing down. Acceleration is a vector quantity, therefore a description of acceleration includes the direction of the force that is changing the speed. In other words, acceleration is the rate of change of velocity in a specific period of time.

Actin: thin muscle filaments.

Adenosine Diphosphate (ADP): a metabolic byproduct of the chemical process that releases energy from ATP [$C_{10}H_{15}N_5O_{10}P_2$].

Adenosine Triphosphate (ATP): a high-energy molecule that stores energy in the muscle cells [$C_{10}H_8N_4O_2NH_2(OH)_2(PO_3H)_3H$].

Aerobic Metabolism: The set of chemical reactions in mitochondria that combine oxygen with glycogen and free fatty acids to produce ATP.

Aerodynamic Drag: a retarding force resulting from the friction between a fluid and an object as the object moves through the fluid.

Aerodynamics: the study of the movement of objects through air.

Analytical Symmetry: a restatement of Newton's third law that says that a moving object in a stationary fluid (like air) is physically equivalent to a stationary object with the fluid moving around it.

Bernoulli Effect: an expression of conservation of energy for compressible fluid flow. In our context, this means that as the velocity of an airstream increases, its pressure decreases and vice versa.

Biochemistry: the study of chemical processes in living organisms.

Biomechanics: a sports science field that applies the laws of physics and mechanics to human performance.

Boundary Layer: the thin layer of fluid adjacent to an object in which molecules of the fluid slow down and tend to directly interact with the object.

Boyle's Law: a relationship between pressure, volume, temperature, and the number of molecules (i.e., the mass of the confined gas) present in the volume. If the number of molecules and the temperature are held constant, the product of the pressure and volume is constant.

Brittle Material: a material that when subjected to a force experiences little or no deformation before breaking.

Calorie: the amount of energy required to raise 1 gram of water 1 degree Celsius.

Center of Mass: the single point on or in any object that can be used to describe that object's response to an external force or torque.

Coefficient of Restitution: the velocity of the ball after striking a surface divided by the velocity of the ball before it strikes the surface.

Compression: the application of an inward force on an object causing it to be reduced in volume.

Conservation of Energy: a principle that states that within any system of objects, energy can neither be created nor destroyed.

Conservation of Momentum: a principle that states that in an interaction between two or more objects (like a foot and a ball), the total momentum of the system is the same before and after the interaction.

Contact Forces: forces which require objects to be in physical contact with one another (like friction, aerodynamic drag).

Drag Coefficient: an empirical factor that relates the drag force on an object to the density of the surrounding fluid (e.g., air) and the square of the velocity of the object.

Elasticity: the tendency of solid materials to return to their original shape after being deformed.

Elastic Material: a material that returns to its original shape after deformation.

Energy: the capacity of a physical system to perform work.

Energy of Motion: (see Kinetic Energy)

Energy of Position: (see Potential Energy)

Fluid Mechanics: the study of the movement of objects through air or other fluids.

Food Calorie (Kilocalorie): 1,000 calories—the amount of energy required to raise 1 kilogram of water 1 degree Celsius.

Force: an influence tending to change the motion of a body or produce motion in a stationary body. Force is equal to the mass of the object, m, times its acceleration, a.

Friction: the force that causes resistance between objects or fluids moving together.

Glycogen: a large molecule produced in the liver that allows the body to store glucose for later use.

Glycolysis: a chain of biochemical reactions that use carbohydrates or sugars in the presence of water to produce ATP.

Hypoglycemia: a low level of glucose in the circulating blood.

Impulse: the product of a force and the duration of the application of the force to an object.

Kilocalorie: (see Food Calorie)

Kinesiology: the study of human motion.

Kinetic Energy: the energy required to move an object of mass, m at a velocity, v. Kinetic Energy is given by $E = 1/2 \, mv^2$.

Magnus Effect: the combination of differential pressure and momentum transfer that causes an object to curve in flight.

Metabolic Energy: energy derived from the food we eat, the water we drink, and the air we breathe.

Moment Arm (also torque arm): the perpendicular distance from the center of the ball to the line of direction of the kick.

Momentum: the tendency of an object to continue moving at a constant speed in a constant direction unless it is disturbed in some way, in accordance with Newton's first law. Momentum equals mass times velocity.

Myosin: thick muscle filaments.

Newton's First Law of Motion: an object will continue to move at a constant velocity unless acted upon by an external force.

Newton's Second Law of Motion: the acceleration of an object is equal to the force that is acting to accelerate it divided by its mass ($a = F/m$).

Newton's Third Law of Motion: when one object exerts a force on another object, the second object exerts an equal force in the opposite direction on the first object.

Non-Contact Forces: forces such as gravity and electromagnetism that act at a distance.

Phosphocreatine: a chemical present in muscle tissue that can regenerate ATP in the cell by donating a phosphate radical to the ADP formed from the original ATP-ADP reaction.

Plastic Material: a material that remains in a deformed condition after a force is applied.

Potential Energy: the energy required to move an object of mass, m to a height, h. In our case, since the force acting against the movement is gravity, Potential Energy is given by *PE = mgh* where *g* is the acceleration of gravity (9.8 m/s²).

Pressure: the force on the inside of the ball (caused by the collisions of air molecules with the inside surface of the ball) divided by the area of the inside of the ball.

Torque: a measure of the force acting on an object that causes that object to rotate.

Turbulence: a condition of airflow in which speed and pressure are changing and in which drag forces are reduced.

Vector: a way to describe physical quantities that have both magnitude and direction, like velocity, acceleration, and force.

Vector Component: the influence that a vector has in a specific direction. Vector components normally form a right triangle. In the case of our soccer analysis, vector components are usually perpendicular to the field and parallel with the field.

Velocity: the amount of distance covered by an object (like a soccer ball) divided by the time required to cover the distance. Velocity is a vector quantity, so a full description also includes its direction of motion. Velocity is measured in units like meters per second (m/s).

Viscosity: a measurement of a fluid's resistance to flow.

Work: the energy required to move an object a certain distance against a given force.

Bibliography

Adair, Robert K. *The Physics of Baseball*. New York: Harper and Row, 1990.

Badiru, Deji. *The Physics of Soccer*. New York: iUniverse, Inc, 2010.

Bray, Ken. *How to Score—Science and the Beautiful Game*. London: Grants Books, 2006.

Briggs, Lyman. "Effect of Spin and Speed on the Lateral Deflection (Curve) of a Baseball; and the Magnus Effect for Smooth Spheres." *American Journal of Physics* 27 (1959): 588–96.

Brown, F. N. M. "See the Wind Blow." *University of Notre Dame Department of Aerospace and Mechanical Engineering Report*. 1971.

Frohlich, Cliff. "Aerodynamic Drag Crisis and its Possible Effect on the Flight of Baseballs." *American Journal of Physics* 52, no. 4, April 1984, 325–34.

Gay, Timothy. *Football Physics—The Science of the Game*. Emmaus, PA: Rodale Press, 2004.

Lees, Adrian, and Lee Nolan. "The Biomechanics of Soccer: A Review." *Journal of Sports Sciences* 16 (1998): 211–34.

Mehta, Rabihandra D. "Aerodynamics of Sports Balls." *Annual Review of Fluid Mechanics* 17 (1985): 151–89.

Morris, Desmond. *The Soccer Tribe*. London: Jonathan Cape, 1981.

Pallis, Jani Macari. "The Fabric of Tennis—The Aerodynamics of Tennis Balls." The Tennis Server. http://www.tennisserver.com/set/set_01_10.html.

Plagenhoef, Stanley. *Patterns of Human Motion*. Englewood Cliffs, NJ: Prentice-Hall, 1971.

Rous, Sir Stanley, and Donald Ford. *A History of the Laws of Association Football*. Zurich: Fédération Internationale de Football Association, 1974.

Shewchenko, N., C. Withnall, M. Keown, R. Gittens, and J. Dvorak. "Heading in Football—Part 2: Biomechanics of Ball Heading and Head Response." *British Journal of Sports Medicine* 39 (Supplement) (2005): i26–i32.

Shinkai, Hironari, Hiroyuki Nunome, Yasuo Ikegami, and Masanori Isokawa. "Ball-Foot Interactions in Impact Phase of the Instep Soccer Kick." *Journal of Sports Science and Medicine*, Supplement 10 (2007): 26–29.

Signy, Dennis. *A Pictoral History of Soccer*. New York: Spring Books, 1968.

Strutt, John (Lord Rayleigh). "On the Irregular Flight of a Tennis Ball." *Messenger of Mathematics* 7 (1877): 14–16.

Tait, Robert Guthrie. "Some Points in the Physics of Golf." *Nature* 42, Issue 1087 (1890): 420–22.

Taylor, John. "Powerpoint Presentation on the History of Soccer." Unpublished. 1995.

Vizard, Frank, ed. *Why a Curveball Curves—The Incredible Science of Sports*. New York: Hearst Books, 2008.

Watts, Robert, and Eric Sawyer. "Aerodynamics of a Knuckle ball." *American Journal of Physics* 41, No. 11 (November 1975): 960–63.

Wesson, John. *The Science of Soccer*. Bristol: Institute of Physics Publishing, 2002.

Williams, Jay H. *The Science Behind Soccer Nutrition*. CreateSpace Publishing, 2012.

Illustration Credits

Index

Page numbers in italic text indicate illustrations.

fatigue, 56
flow separation, 39, 41
fluid mechanics, 22, 87
follow-through, 36, 36n
food energy. *See* energy, metabolic
force, 12–14, 16–17, 21, 46, 87–89;
    contact, 14, 17 19, 86; electromagnetic,
    18, 88; gravitational, 18; non-contact,
    17, 88; nuclear, 18
Fountain Valley School, 49
free kick, 1, *1*, 12, *35*, 61, *62*
friction, 17–18, 24, 32, 87

Gaelic Athletic Association, 8, *8*
Galen, *10*, 11
Galilei, Galileo, 9
glucose, 53–55, 56
glycogen, 54, *54*, 56, 87
glycolysis, 55, 87
goal kick, 12
gravity, 15–17, 70, 88

hamstring muscle, 45, 58
harpastron, 4–5
harpastum, 3, 5, *8*
heading (the ball), 3, 46, *47*, 59, *61*; diving,
    47–48; glancing, 47–48; standing or
    jumping, *47*
Henry VIII (king of England), 6
hip flexor muscle, 45
history of soccer, 3–9
hurling, 8
hydration, 54–55
hypoglycemia, 56, 87

impulse, 14, 52, 52n, 87
Ireland, 8
Italy, 5–6

James IV (king of England), 6
Julius Caesar, 5

kemari, 3, *8*
kicking (the ball), 3, 14–18, 37–38, 40, 42,
    45, *45*, 58, 60, *60*, 75–77
kickoff, 17
kinesiology, 10–11, 87
knuckleball, 41
koura, 3

London, 6
Lord Raleigh. *See* Strutt, John

Magnus, Heinrich, 9, 40
Magnus effect, 39, *39*, 40, *40*, *41*, 42–43,
    61–62, 72–74, 88
mass, 12, 12n, 15–17, 87–88
moment arm, 23, *23*, 88
momentum, 15, 21n, 88; conservation of,
    21, 39, 86
myosin, 55, 88

Newton, Sir Isaac, 9, 10, *16*
Newton's laws of motion, 16–17, 39;
    Newton's first law, 16, 21, 88; Newton's
    second law, 16, 88; Newton's third law,
    16, *17*, 19, 85, 88
nucleus (of an atom), 18–19

parrying (by goalkeeper), 52, *53*
perspiration, 55
phosphocreatine, 55, 88
plastic material, 28, 88
pressure, 16–17, 20, 29, *30*, 38–39, 71,
    88–89
Puritans, 6
pusuckquakkohowog, 5

quadriceps muscles, 46

referee, 1, 61
Richard II (king of England), 6
rolling, 20
Royal Navy, 6–7, *7*
rugby (game), 8
Rugby (school), 8
Rugby League, 8
Rugby Union, *8*
rules of soccer, 7–8

Shrove Tuesday games, 3–5, *6*, *8*
skidding, 32, *32*, 33
soule, 5, *8*
spinning, 3, 22, *23*, 36–38, *39*
Strutt, John (Lord Raleigh), 10, *10*
sugar. *See* glucose
symmetry of forces. *See* Newton's laws of
    motion, Newton's third law